T0195308

Inscriptions of Nature

Inscriptions of Nature

Geology and the Naturalization of Antiquity

PRATIK CHAKRABARTI

Johns Hopkins University Press *Baltimore*

Printed in the United States of America on acid-free paper
9 8 7 6 5 4 3 2

Johns Hopkins University Press
2715 North Charles Street
Baltimore, Maryland 21218-4363
www.press.jhu.edu

Library of Congress Cataloging-in-Publication Data

Names: Chakrabarati, Pratik, author.
Title: Inscriptions of nature : geology and the naturalization of antiquity / Pratik Chakrabarti.
Description: Baltimore : Johns Hopkins University Press, 2020. | Includes bibliographical references and index.
Identifiers: LCCN 2019056149 | ISBN 9781421438740 (hardcover) | ISBN 9781421438757 (ebook)
Subjects: LCSH: Archaeological geology—India—Gondwana. | Geology—Methodology. | Gondwana (India)—Historical geography. | Gondwana (India)—Environmental conditions. | Gondwana (India)—Antiquities. | India—Historiography.
Classification: LCC DS485.G647 C43 2020 | DDC 934/.3—dc23
LC record available at https://lccn.loc.gov/2019056149

A catalog record for this book is available from the British Library.

Special discounts are available for bulk purchases of this book. For more information, please contact Special Sales at specialsales@jh.edu.

Johns Hopkins University Press uses environmentally friendly book materials, including recycled text paper that is composed of at least 30 percent post-consumer waste, whenever possible.

For Baba

In my dreams, I see his boat sail across the river from this world to the other.

CONTENTS

ILLUSTRATIONS

I am grateful to the Leverhulme trust for generously funding the project "An Antique Land; Geology, Philology and the Making of the Indian Subcontinent, 1830–1920," which I held at the University of Kent and University of Manchester. This enabled my research associate Joydeep Sen and me to undertake archival research and coauthor one of the articles from the project. The project benefited immensely from Joydeep's diligence, incisive observations, and intellectual input. I have thoroughly enjoyed our discussions on fossils, sacred geography, mythemes, and of course sports. I was also fortunate to work on this project with my graduate student Cam Sharp Jones.

It takes a community of scholars and friends to raise a book. I am deeply indebted to those who have helped me in various ways, often without asking, and whenever I needed. In particular, I would like to thank Sangeeta Dasgupta, Padmanabh Samarendra, and Gaurab Datta for being true friends. I cherish your friendship, your unquestioning loyalty, and your unwavering support over the years and particularly during the writing of this book.

I am grateful to Sarah Irving, Saul Dubow, Aninidita Ghosh, Vladimir Jankovic and Ian Burney (both my colleagues at CHSTM), Projit Mukharji, Sunil Kumar, Linda Andersson Burnett, Sanjukta Das Gupta, Kavita Shivaramakrishnan, and Graham Mooney for making valuable suggestions about the content and plan of the book. A special thanks to Simona Boscani Leoni for inviting me to the wonderful workshop ("Mapping the Territory: Exploring People and Nature, 1700–1830," at the Center for Global Studies, Universität Bern, September 2017). It gave me the opportunity to present my paper on the canal of Zabita Khan in its formative stage, and the discussions there helped me significantly to develop the chapter. Parts of the research for the project were published in articles in *Modern Asian Studies* (jointly with Joydeep Sen) and *Past & Present*.

On several occasions, I have turned to Mick Worboys, Neeladri Bhattacharya, Mark Harrison, and Alison Bashford for their advice and suggestions and they have been extremely generous with their time and assistance. I am grateful to Shivi (K. Sivaramakrishnan) for inviting me to the MacMillan Center at Yale to present my paper and conduct research at the Yale University Archives.

The staff at the University of Manchester Library, National Library of Scotland, the British Library, the National Army Museum, and the American Museum of Natural History were extremely helpful in facilitating the copies of the several images and the permission to use these in the book. Sue and Bill

Peppé helped me with one of the photographs used in the book. I would like to thank Matt McAdam of Johns Hopkins University Press for his faith in the book. Juliana McCarthy, William Krause, and Hilary S. Jacqmin saw the book through production smoothly. Carrie Watterson's meticulous and incisive editing improved the text vastly. Finally, I am grateful to my colleagues at CHSTM, Manchester, for welcoming me to the institution. It is an absolute pleasure to work among colleagues who are so brilliant in their own scholarship and genuinely interested in and supportive of each other's work.

In many ways, this book took me back to some of the roots of my intellectual journey. Here I have written about the only city I have known as my home: the evocative, layered, and labyrinthine city of Delhi, which has defined me as a historian. In the same breath, I wish to acknowledge the role of my alma mater, Jawaharlal Nehru University, in shaping my critical and creative acumen.

In Delhi, at JNU, I met Nandini. She has remained the one constant source of conviction and courage in my life ever since. Her love, perspicacity, and patience are the only reasons I am able to sustain my academic life. I owe this book to her.

I am grateful to my sister for being the source of strength in times of tragedy and to my mother for making it always possible for us to pursue our dreams and ambitions.

My father passed away quite suddenly, as I was finishing the book. I feel his love, and I miss him every day. The words I write will never suffice for those that have remained unsaid between us. I have tried to emulate his example of hard work, generosity, compassion, and conscientiousness in my own life and work. I dedicate this book to him.

Inscriptions of Nature

Past Unlimited

The true wonders of the country are under the surface.

Gazetteer of the Central Provinces of India, 1870

In the nineteenth century, geologists, archaeologists, Orientalists, engineers, and urban planners began digging the earth like never before. They were digging for sanitation, transport, irrigation, minerals, and antiquity. Geology emerged as the key theme in this visceral encounter with the earth. It, therefore, provides insights into the formation of states, the mining of natural resources, and the creation of national topographies. It also unearths imaginations of deep history. The geological imagination of India, its landscape, people, past, and destiny drives the narrative of this book. Yet geologists were not alone in this pursuit of terrestrial resources and antiquity, nor was geology a singular discipline. Sometimes digging to create underground sanitation systems revealed human remains, excavation of ancient cities unearthed prehistoric fossils, or cutting channels for canals uncovered buried cities. Besides the earth, geologists, ethnologists, archaeologists, and missionaries were also digging into ancient texts and genealogies and delving into the lives and bodies of indigenous populations, their myths, legends, and pasts. One pursuit was imbricated with another in this encounter with the earth and its inhabitants and between the past, present, and future.

In these layered experiences of antiquity, several notions (and disciplines) of the past coexisted and informed each other. Hugh Falconer, one of the prominent geologists of the nineteenth century, believed that fossils were the "monuments and inscriptions constructed by nature."[1] Falconer's dual metaphor indicates the concurrent pursuit of antiquity in natural and cultural landscapes. This duality of the real and metaphorical digging of the earth naturalized the past, indelibly *inscribing* the nineteenth-century study and imagination of antiquity in the deep history of nature, landscape, and people.

To clarify what is meant by "naturalized" throughout this book, consider the distinction that emerged in the nineteenth century between the "historical" and the "geological." In the 1830s, based on his paleontological discoveries of human remains, the French polymath Paul Tournal proposed the term "historical" be used only for the period of recorded human history, to distinguish it from the geological period, which he called *antéhistorique*, comprising a both human and geological history devoid of decipherable records.[2] While the historical period could be studied by conventional means such as texts, inscriptions, and numismatics, Tournal contended that geological and paleontological evidence was critical for studying the *antéhistorique*, which later came to be referred to in English as "prehistory."[3] Prehistory was therefore defined by nature, a form of antiquity that could be traced only in the strata, fossils, and flint stones. Tournal's distinction between the historical and geological or natural has persisted at an epistemological level in the study of deep history. In this book, "naturalized" refers to the emerging deep history of nature, which became a pervasive category of antiquarian thinking for both the geological and historical periods.

Although in this book I use conventional expressions—that is, the historical and the geological or the natural—I also highlight the essential interplay between them in the nineteenth century. I show that in geological fieldwork, ethnographic research, and the study of myths, the historical and the prehistorical overlapped. Ethnographers used geological evidence to study contemporary tribes or tribal myths, while geologists referred to written texts or oral traditions to study prehistoric rock formations and fossils. It is through exploration of this interplay that my main argument emerges: that the naturalistic frame, with its hidden historical transcript, formed a foundation of historical imagination per se.

I explore the historical and the natural in the construction of the Yamuna (Doab) Canal in northern India in the early nineteenth century. The digging for the canal unearthed ancient canal networks, lost riverbeds, traces of mythological rivers, and prehistoric fossils that inspired the formative ideas of Indian antiquity, prehistoric geography, and human evolution in the Indian subcontinent. The terrestrial excavations were complemented with the study of local myths, historical texts, and ethnological accounts. The historical antiquarianism around the medieval canal system and other archaeological remains in the Doab region were infused with geological antiquity. Consequently, the Indo-Gangetic Plain appeared as the bedrock of Indian civilization and antiquity.

I also analyze the dense forests, hills, and ravines of central India where ethnological studies among tribes and geological investigations of some of the oldest rock formations of the Indian subcontinent simultaneously shaped ideas of Indian aboriginality and prehistory. At the same time, the natural resources of central India provided vital wealth for the nineteenth-century colonial state in the form of coal and cotton, often extracted by tribal labor. Through the geological, anthropological, and social history of central India, I identify the colonial appropriation of India's deep antiquity to the colonial present to consider how themes of human evolution, mythological geographies, Aryanism, and tribal aboriginality defined Indian antiquity.

Two main inquiries run through the course of the book. First, to understand how along the banks of the Yamuna and Ganga, among the Himalayan crests, in the deep ravines and forests of central India, and through their associated myths and legends, historical imagination came to be infused with the deep history of nature. The second aim of the book is to show that because of the fusion of natural and historical imaginations, the study of Indian antiquity was not often confined to specific disciplinary boundaries.

Nature has been part of historical imagination since antiquity. Jean Bodin and Montesquieu saw societies embedded in nature and wished social laws to align with natural laws.[4] In India as well, premodern architectural and antiquarian sensibilities were often infused with impressions of nature.[5] Ideas of race also emerged from collective reflections on nature, climate, and environment from the sixteenth century. However, these associations between nature and antiquity were made *without* the intense encounters with the earth and its inhabitants that took place in the nineteenth century. The earthly experiences of this period led to a distinct, intense, and deterministic alignment of natural and historical imagination, which established the naturalistic frame of antiquity.

That frame is evident in some of the most prominent literature that connects human history with nature. The finest exposition is Fernand Braudel's geohistorical account of the Mediterranean world. Braudel placed the *geohistoire* of the mountains, deserts, and seas at the foundations of human history.[6] A more recent elucidation is Sunil Amrith's study of the migration of people across the vast Monsoon Asia region, a history of individuals whose lives were defined by the climatic patterns of the Bay of Bengal. Amrith adopts the Braudelian frame of seeing the geohistorical setting as the foundation on which social and cultural history unfolds. While Braudel believed that "mountains come first," Amrith explains that the "monsoon sustains life in the Bay."[7] In the same vein, William Beinart has shown that the Karoo region set the "rhythms of life" in the veld of South Africa.[8] Chetan Singh argues that the ecology of western Himalaya virtually "governs" human life there.[9]

Environmental history as a distinct discipline emerged in the twentieth century. Charles Pergameni was the first to formally define "historical geography" as a study of human geography in the 1940s.[10] As a recent book has shown, the idea of the "environment" itself, as different from the idea of nature, developed in the post–World War II era of awareness of environmental degeneration, particularly evident in William Vogt's *Road to Survival*, published in 1948.[11] Yet both environmental history and historical geography draw from the deep history of nature that emerged in the nineteenth century. They have incorporated this naturalistic frame as an essential way to understand long-term human relationships with the environment, eventual ecological degradation, and ways of restoring a more equitable relationship between nature and humanity.[12] Donald Worster analyzed the emergence of the modern ecological thinking in Western intellectual traditions through the contradictory modes of Romanticism, industrialism, and the mechanization of the view of nature.[13] He sees the Romantic approach to nature as ecologically holistic, in contrast to that shaped by the modern exploitation of nature that created, for example, the "dust bowl" of the southern plains of the United States.[14] Simon Schama has shown that the history of the three key features of nature—the rocks, waters, and woods—are inscribed with human culture, memory, and legends, which too carry traces of this history of nature. He shows that these were also lost by human action. The primeval quality of the great forests of Białowieża in central and eastern Europe, for example, was shattered first by the Aryan puritanism of Nazi Germany and then by the statist forestry of the Soviet Union.[15] Both Schama and Worster see nature as essentially pristine and Romantic, especially in its deep primordial state.

These are compelling narratives. They appear particularly convincing *because* they link history with deep nature, and thereby the past appears more embedded and anchored. The deep naturalism that underpins the works of Schama, Braudel, Amrith, and Worster emerged out of the nineteenth-century explorations of deep time.[16] These associations also appear fundamental to the future of historical writing. Participants in a recent roundtable discussion on history and biology declared, "Human history and 'natural' history are intertwined. Neither can be understood without the other."[17] This proposition specified how historians could incorporate, for instance, biological tools such as epigenetics into their analysis. Yet such conceptual positioning precludes the possibility of seeing nature itself as a matter of human imagination; what these people crossed or inhabited in the Mediterranean, the Karoo, or the Bay of Bengal was not just the sea, the mountains, and the deserts but their imaginations of these natural worlds and their frontiers. This is not to invert the formulation and suggest a cultural construction of nature but to trace the genesis of this geohistorical thinking.

At the heart of this story is the European discovery of deep time. The discovery of plant, animal, and human fossils and flint tools placed the study of the history of the earth and of human antiquity beyond the timescale of textual or religious traditions. Martin Rudwick in his grand narrative has traced the emergence of European deep historical thinking from the eighteenth century. Rudwick shows that European savants such as Horace Bénédict de Saussure, Georges Cuvier, William Hamilton, and Charles Lyell derived ideas of the deep past from complex sources of eighteenth-century antiquarianism. In the nineteenth century, geology provided a new paradigm for the modern understanding of time.[18]

This book marks three points of departure from Rudwick's thesis and the general narrative of the history of European geology. First, it traces the discovery of deep time in colonial history, which acquired complex meanings in India, Australia, and South Africa, in its encounters with colonial antiquarianism, Orientalism, exploitation of natural resources, and the study of aboriginality and tribal mythologies. Deep time and deep history are important themes in the scholarship on the relationships between colonialism, nature, and human antiquity.[19] In Australia, deep time has become a critical theme in writing Aboriginal relationships with the Australian landscape.[20] In this literature, historians, archaeologists, and anthropologists combined the European geological notion of deep time with aboriginal ideas of time.[21] In India, there is no similar historical work on deep time. Prathama Banerjee has examined the various notions of time that functioned within the colonial imagination of primitivism in India.[22] Thomas Trautmann has highlighted the conflict between and conflation of Indian (derived chiefly from Hindu/Sanskrit texts) and European notions of time in the eighteenth and early nineteenth centuries.[23] He has also analyzed the contemporaneity of the deep past in Indian modernity.[24] In both instances, he has worked on the "historical" deep past starting from the Indus Valley civilization to ancient India and does not refer to the making of prehistoric geological time. My examination of the interplay between colonialism, antiquarianism, and prehistory in India endeavors to fill this gap.

Second, Rudwick explains the European discovery of deep geological time, not its infusion within historical imagination per se. To that extent, Rudwick's work is confined to the geological narrative of deep time, despite the complex lineages of that thinking that it traces. To give one example, for the subtitle of his book *Bursting the Limits*, he borrows the term "geohistory" (which forms the central theme of his work), from modern earth scientists, who use it as a purely geological category, in contradistinction to Braudel's and other historians' earlier and much broader use.[25] In fact, Braudel's book, which is widely regarded as one of the finest expositions of geohistory, is not even included in the bibliography of Rudwick's books on geohistory. While Rudwick's work remains confined to the geological narrative, this book does not. A wider focus allows me to investigate not just the making of the Indian imagination of geological deep time but also the *politics* of deep time. I, therefore, examine how deep history became pervasive and dominant in the imagination of the past in general.

Third, historians have appreciated that with the emergence of deep history, nature became a historical landscape. Rudwick has shown that the geological explorations of the eighteenth and nineteenth centuries "historiciz[ed] the earth."[26] Geologists studied the earth (its strata, fossils, craters, rocks, and mountains) like a historical text to understand its past. Vybarr Cregan-Reid has made the point that in the nineteenth century, geology, in fact, replaced the existing anthropocentric and narrative forms of antiquity. Because deep time was "unnarratable" in anthropocentric terms, it became the new narrative of antiquity.[27]

This book suggests that deep time in fact provided a new narrative to anthropocentric histories, in which the deep history of rocks, landscapes, and rivers conferred new meanings upon historical antiquity. As we shall see, in the nineteenth century, erstwhile textual references to antiquity were placed alongside "field" notes, enmeshing Tournal's historical and the prehistorical. Mythical rivers such as the Saraswati, references to which were found in Hindu, Persian, and Arabic texts, were searched for in the dried riverbeds of the Indo-Gangetic Plain. Similarly, the mythological monkey army of the *Ramayana* became anthropomorphized into the prehistoric races of central India.

To put it simply, while the general trend of the history of geology has been to study the historicization of the earth, this book explores the naturalization of history. The past, including human past, appeared more authentic when it was inscribed in deep natural history, landscape, and ecology. The binary between the ecological and the materialistic, which informs Schama's and Worster's work, appears less striking when we understand this process. The mythical Arcadia, for instance, became a tangible geographical entity "under glass" in the London Zoo not only because of the "pseudo-naturalization" of the zoo as Schama suggests but also because of the naturalization of antiquity in the nineteenth century.[28]

Chapter 1 presents a key moment in the emergence of the new naturalized narrative of antiquity. It begins with the digging of the Doab Canal in the 1820s and analyzes how a colonial engineering project becomes simultaneously an archaeological and a geological endeavor. The canal is the first site of examination in this book where history became naturalized. In the courses of the excavation, ancient canals appeared indistinguishable from old riverbeds and mythical rivers, and the line between monument and terrain, or the "natural" and the "historical," became imperceptible. In the process, the landscape, the legends, and the monuments became part of this colonial antiquarianism. Chapter 2 then shows how Orientalist accounts of Indian antiquity, derived from Hindu texts and inscriptions, came to reside within this emerging geological narrative. The fossils found in the Siwalik foothills of the Himalayas and the Gangetic Plain defined theories of not only the emergence of primeval humans in India but also the formation of the entire subcontinent. Crucially, these theories of human evolution drew not just from the discovery of fossils but also from a range of Orientalist and Indological traditions that simultaneously deliberated upon the origin of human civilization in India.

Chapter 3 explores how the Orientalist reading of Hindu mythological traditions, such as the Puranas, as historical texts acquired geomythological connotations in the nineteenth century. Through their geological redefinition, Hindu myths and sacred geographical ideas emerged as more naturalistic and enduring entities in the Hindu imagination of India. Chapters 4 and 5 investigate the links between aboriginality, primitivism, and prehistory. Chapter 4 shows how indigenous populations in South Asia were seen as remnants of prehistoric races and how subsequently their histories, lifestyles, cultures, and myths were viewed through paleontological and geological frames. The final chapter interrogates how cultural and ethnological ideas of primitivism, usually applied to the marginal populations of the global south, acquired geological meanings, particularly in the imagination of the prehistoric southern continent of Gondwanaland. It shows that Gondwanaland took shape as a primeval land when it became the repository of various colonial imaginations of tribal and geological prehistory in the nineteenth century.

An understanding of the naturalized frame of history allows us to reflect critically on the literature of the "Anthropocene," which has introduced a new urgency to naturalizing history. Emerging from the debates on climate change, the literature suggests that there has been a fundamental change in human agency vis-à-vis nature and that, by making substantial changes to its ecologies, humans have become a "geological force," "changing the most basic physical processes of the earth."[29] This view has collapsed the "age-old humanist distinction between natural history and human history."[30] Therefore, the literature suggests that we need to align existing forms of social, cultural, and environmental history with scientific understandings of a geological past and think in terms of planetary consciousness to piece together the contextual fragments of a grand narrative whose conclusions seem to have been already written.

Since the literature emerges from a great urgency forewarned by scientists—global warming—with the expectation that historians react to it, it privileges scientific explanations of the past over the historical. For historians the question is, in writing the histories of species, should the history of science simulate science or interrogate it? The specter of global warming itself should not prevent us from questioning the sanctity of its historicity. It is not simply a matter of whether scientific facts are accurate but also whether the scientific narratives of the past should define the historical narrative and how such scientific narratives of antiquity became dominant and persuasive. Historians of science need to continue to raise questions about science and about antiquity that geologists, paleoanthropologists, evolutionary biologists, and geneticists do not ask or contemplate or even appreciate. For example, what kinds of engagements with nature made the imagination of deep past (which in essence is intensive and, in some respects, absolute knowledge of nature) possible? How did this knowledge evolve in the nineteenth century, at the same time that vast natural resources, particularly in the colonies, were being encroached upon and exploited, which in turn enabled the fossil fuel economy to emerge? Is the deep past complicit in the Western and colonial appropriation of global nature, time, myths, and commodities?[31] It is crucial to revisit the moment when that link between history and deep nature was created. It was in the nineteenth century that the "age-old" divide between the natural and the historical that Dipesh Chakrabarty refers to was first transgressed. The history of the naturalization of antiquity lies in this transgression. It allows us to understand why the Anthropocene, a product of such nineteenth-century thinking, appears to be a holistic and comprehensive historical concept and presents the prospect of naturalizing history all over again.

As this book seeks to elucidate how the notion of deep time became a pervasive category of understanding the past, it also traces some of the unique characteristics of Indian antiquity as constituted in the nineteenth century. The Indological interpretation of Hindu antiquity evolved in the nineteenth century at the time of, and in close interaction with, the British discovery of deep time in the Indian subcontinent. Therefore, Hindu antiquity readily incorporated notions of deep time. To explain this process, I use the concept of "past unlimited," borrowing from Chakrabarty's phrase "politics unlimited." Chakrabarty uses the term to argue that the marginalized and the oppressed communities in India have to adopt every means at hand to fight the established political system, and therefore their political domain, in principle, has no limits. This is quite distinct, Chakrabarty suggests, from the politics of Aboriginal rights in Australia, which has had to adhere to the strict academic disciplinary norms of history, archaeology, prehistory, and anthropology.[32] It is, of course, necessary to point out that in India such an unlimited political scope is available not just to the marginalized but to elite Hindus as well. They too have adopted ideas from various academic disciplines and combined them with myths and folklore to assert their versions of truth, antiquity, and politics. Therefore the term "unlimited" as used in this book is not necessarily only derived from a sense and experience of marginalization. Rather, it reflects the general lack of sanctity of academic disciplines and sites of antiquity in the study of the past in Indian political, intellectual, and social life.

The phrase "past unlimited" is useful in problematizing deep time in India. With it, I argue that in India the pursuit of the deep past crossed two thresholds. The first was the threshold between disciplines, such as history, anthropology, archaeology, mythology, and geology. The second was between natural and cultural landscapes, which tapped the unlimited possibilities of antiquity. Hindu antiquarianism therefore uninhibitedly blended science, history, and deep time with myths, and it imagined India as a deeply Hindu land. These processes of naturalized antiquarianism of the nineteenth century took place not just in India but also in Africa and Australia, and the various chapters draw examples from these regions as well. However, the main narrative of the book is based in the Indian subcontinent.

Past Unlimited: Myths and the Deepening of History

How did deep time establish itself within Indian natural and cultural landscapes? One clue is in the enduring relationship between myths and the deep history of nature that formed in nineteenth-century India. Myths became entrenched in public culture and memory when they were linked to the geological features and events of the subcontinent. At a time when geologists in

Europe constructed deep history increasingly on secular and geological grounds, shedding their biblical interpretations, in India geologists continued to refer to allegorical texts, such as the Puranas and their associated myths.[33] These narratives remained central to constructing the imagination of the geological evolution of the Indian landscape and its human and nonhuman inhabitants.

On the one hand, deep naturalism provided myths with new cultural potency. On the other, myths provided deep time with its cultural and social settings in India. Chapters 2 and 3 show that Indian fossils, along with Indian landscapes and rocks, were ascribed Hindu mythological meanings as they were analyzed along with Orientalist reading of the Puranas. Early Orientalist scholarship, in its search for the Judeo-Christian roots of Hinduism, studied Hindu mythological texts for references to Mosaic histories and in the process ascribed powerful sacred imageries—derived simultaneously from biblical and Hindu lineages—to Indian fossils. This led to the retention of religious ideas within Indian geology and deep time and the consilience of science and mythology, long after such ideas were discarded in Europe. The deep past in India is therefore not just a distant entity hidden in caves or the depths of the earth. It lies in the cultural imaginations of the nation, in tribal ideas of their homeland, and the unlimited possibilities of the living past. It was shaped by simultaneous geological and mythological experiences in the reading of classical texts and along the banks of the Yamuna and on the hilltops of Gondwana.[34]

The history of the relationship between geology and myths has been written predominantly from European and biblical perspectives. Historians have explained how in the nineteenth century, geology replaced the biblical imagery of the deluge and the Adamic creation of the world.[35] In Europe, the emergence of the idea of deep time gradually reconfigured biblical time and history, and the Genesis story was redefined, especially with ideas of pre-Adamite hominids, to suggest that primitive ancestors of Adam had populated the earth before Adam and Eve.[36] These mythological ideas were gradually abandoned in favor of a more secular vision of earth history.

Recently, scholars have shown that non-European cultural contexts and religious metaphors played crucial roles in redefining geohistory.[37] At the same time, they have followed the significant traces of Judeo-Christian themes in the colonial and postcolonial worlds. The Orientalist scholar William Jones believed that all great civilizations had common roots and that the discovery of a common place of origin in the Middle East was a vindication of the book of Genesis.[38] The motive of his study of Sanskrit and Hinduism was his desire to vindicate the Mosaic history of the primitive world. In Africa, Judeo-Christian geomyths engendered new political identities among Tutsis and Hutus, ultimately leading to the Rwandan genocide.[39] These geomyths took deep root in Asia and Africa as they blended with existing cultural and antiquarian ideas. How the infusion of deep history provided myths with enduring historical and political legacies in India is the central question of this study.

Myths and the deep past are significant sites for establishing the authenticity of various historical debates in modern India. They are also sites of political disputes about the validity of various traditions. Paleoanthropologists, archaeologists, and geologists continually resort to ancient Puranic myths when engaging with geological and archaeological questions or vice versa. At the same time, scientists have searched for the mythical river Saraswati in the desert tracts of northwestern India and have deliberated upon whether the chain of limestone shoals between the southern tip of the Indian subcontinent and Sri Lanka was the Rama Setu (the mythical bridge constructed by Lord Rama to reach Lanka to rescue his wife Sita).[40] The latter is also known as Adam's Bridge because of its biblical and Islamic connotations as one of the sites of Adam's descent to earth in Asia. In the twenty-first century, DNA research, archaeology, and linguistics have combined to validate Hindu theories of India being the home of the Aryans in the nineteenth century.[41] Myths have also informed historical and geological narratives. Sumathi Ramaswamy traces memory, identity, and loss through the reenactments of the imaginary land of Lemuria among the nationalists of southern India. She shows how the deep past of Lemuria became a living entity at a critical juncture of Indian modernity, besieged as it was then by a sense of a lost past.[42]

Scholars have deliberated at length on the relationship between myths and history in India, because most Indian historical traditions appear to be mythological. Ashis Nandy stresses the need to see Indian mythological traditions as historical, rather than as purely allegorical and fictional. For him, the Puranas represent the true tradition of historical writing in India.[43] This, in fact, recapitulates the eighteenth-century Orientalist readings of these texts as historical literature shrouded in allegorical representations.[44] Others have similarly suggested that the discovery of the authentic traditions of the "Indian" past and the concomitant redefinition of history as a European project requires a greater appreciation of Indian sacred or mythological traditions.[45] These are significant interventions, but they essentialize the proposition that while the West wrote history, the Eastern traditions of antiquity were vested in myths, which then require interpretation into modern historical terms. This proposition precludes investigating why and how myths appear historical and as essential parts of Indian antiquity in the first place. More importantly, it engenders the problem of seeing and validating myths as historical.

There was a particular moment in the history of postcolonial India when questions of myths, deep history, and science became critical to the identity of the nation. The infamous Ram Janmabhoomi (birthplace of Rama) crisis of the 1990s centered on claims made by certain Hindus that the site where the Babri Masjid (a sixteenth-century mosque) once stood in the city of Ayodhya in northern India was the birthplace of the Hindu god Rama. Debate ensued among historians, archaeologists, and Indologists whether that was indeed the case and whether a Hindu temple that commemorated the birthplace of Rama had been demolished to build the mosque. Political events escalated, and during a rally on December 6, 1992, the mosque was demolished by a mob. These events have understandably generated a great range of political and historical literature. The debate has concentrated on whether the claims are true, whether the site is indeed the birthplace of Rama, and whether thereby mythology can be transcribed into historical evidence.[46]

Here I do not address the authenticity of such claims. Rather, I seek to understand why archaeology and myths have coexisted in postcolonial India. Historians have understandably placed their faith in scientific methods, particularly in archaeology, to resist the narrativization of myths as historical truth. Tapati Guha-Thakurta argues that during that dispute, not only the site but also the science of archaeology itself were under threat.[47] The integrity of objective knowledge was challenged, as the Hindu Right indiscriminately fused archaeology with scriptural texts, epics, and testimonies of travelers and made selective use of archival documents. The need was, she persuasively argued, to return Ayodhya to archaeology and every piece of archaeological evidence to its respective strata or stratigraphic analysis to separate each period from the other.[48]

The question is, can archaeology provide that elusive authenticity, particularly when it is implicated in the reinvention of myths as history? The problem is not just in verifying whether specific sites such as the mosque or the dead riverbed were, in fact, the places where myths identified them to be but also in how archaeological sites reinvented myths as historical subjects in the nineteenth century. Neeladri Bhattacharya has shown that myths have a closer relationship with secular historical traditions than is generally appreciated. The Ram Janmabhoomi controversy generated competing notions of pasts and myths. While secular historians could reject the Hindu Right's politics, they were unable to reject their notions of the past entirely, which remained popular.[49] Although historians seek to write secular versions of the past, in doing so, they depend on mythical texts and traditions. To understand how myths have remained powerful in Indian society, Bhattacharya refers to the process of "mythification" through which myths are rendered both true and timeless. Elements of the past that reside in popular memory are amalgamated with a modern political discourse, and the new political discourse legitimizes itself through known and familiar traditions.[50] It is possible to draw parallels between Bhattacharya's analysis and Edward W. Said's study of Jerusalem, a site layered with intense religious meanings where collective memory is constantly shaped by myths and modern politics.[51]

The project of this book is to decipher not just the political processes of mythification but how modern archaeological and geological interventions caused myths to appear as historical facts. These developments, I argue, became particularly evident in the nineteenth century, when myths were linked to deep histories. In this period, the birthplace of Rama was identified as a historical site as much as the monkey armies of the same epic were interpreted as aboriginal tribes of India.[52] Myths, as they were linked to specific geographical sites or geological phenomena, assumed more physical and tangible shapes. In other words, myths seemed real when they were inscribed onto the landscape, geological events, and thereby the deep past of the nation.

Why did this occur in the nineteenth century? Indian mythological traditions such as the Puranas were infused into the modern historical imagination at the same time that the sciences of deep time searched for traces of geological and archaeological pasts in Indian landscapes. Beginning with Jones in the late eighteenth to Horace Hayman Wilson in the mid-nineteenth century, Orientalist scholars referred to the Puranas as the allegorical basis of the Indian past. Jones's reading of the allegorical meanings of Puranic texts for evidence of the biblical deluge, for instance, established the possibility of Hindus later reading such texts for diverse geohistorical meanings, for instance, transmuting biblical geographical motifs into propositions of Indo-Aryanism.[53] Meanwhile, British geologists, anthropologists, and archaeologists consulted the Puranas, *Ramayana*, *Mahabharata*, and Persian and Arabic texts for depictions of the geohistorical accounts of Indian fossils and tribes and for explanations of seismic changes in the courses of Indo-Gangetic rivers or the formation of Himalayas. In the process, I suggest, myths were established as part of the deep history of India, and Indian antiquity itself was established through a fusion of history, myths, and deep history. To that extent, myths and history followed the same trajectory of gaining authenticity through their association with the deep history of nature, which is why it is difficult, in such instances, to distinguish historical facts from myths.

Seeing the Past

This leads us to the question of the historical framing of deep time. Historians are usually reticent in engaging with deep history, mainly because their usual methods and tools, such as the examination of textual, epigraphic, and oral sources, seem to have little relevance on that timescale. When they *have* joined collaborative projects with anthropologists, archaeologists, and geologists in writing deep history, they have provided insights into the long history of linguistic developments, cultural practices, and dietary habits, as Daniel Lord Smail and Trautmann have done in the multiauthored volume *Deep History*. These authors also drew insights from medieval and early modern histories and projected them back into the deep past.[54] While they have created opportunities for historians to contribute to the analyses of the deep past, they have not necessarily analyzed how historical methods can engage critically with conceptions of deep time. Their approach tends to confirm the geological, anthropological, and archaeological settings of deep time rather than critique them.

One of the ways historians can creatively intervene in deep history is by problematizing the role of nature—by exploring how nature, which appeared as *the* unbroken link between the deep past and present, defined prehistory. To initiate this process, I ask a simple question: What did the geologists, ethnologists, and archaeologists see? Geology, one of the main proponents of deep time, is itself a historical discipline, as it is an imagination of the past forms of the earth. Yet all that geologists see is the earth or the natural world in its present form, including its rocks, fossils, and inhabitants. This earth is neither prehistoric nor is it primitive. It is in these modern rocks and living forms that the geologist sees traces of prehistory. To that extent, geology is not just an exploration of deep time. It is an analysis of the past as an analogy of the present.[55] In such an undertaking, nature forms the vital clue, as it is through the naturalization of the present that the geologist uncovers the deep past.

In such an initiative, geology in the nineteenth century was not dissimilar to anthropology or prehistoric archaeology. As we will see in chapter 4, in the nineteenth century, anthropologists deliberated upon human prehistory through the observation of contemporary tribes and indigenous populations as relics of prehistoric humans. In doing so, geological and anthropological methodologies often overlapped. In the case of Gondwanaland, geologists drew from anthropological insights of the study of Gond tribes and vice versa. These disciplinary overlaps are compounded by the problem of chronology. As British explorers excavated precolonial canals to build the colonial irrigation system in the Ganga-Yamuna basin or extracted coal and diamond from mines in central India, they straddled modernity, medieval history, classical antiquity, and the deep past. The different disciplines eventually allocated them to geological, archaeological, and anthropological timescales and thereby linearized what they had in fact confronted simultaneously. However, this book captures them at this moment of disciplinary innocence when they remained tangled in a rather chaotic mess.

This moment is fundamental to my inquiry, as it presents two possibilities. First, it problematizes how the present shaped ideas of deep time through nature and continues to do so. Second, this naturalization of antiquity needs to be situated at the confluence of intellectual disciplines such as geology, archaeology, and anthropology. A deep synthesis of ideas took place between these disciplines in the nineteenth century, and they collectively invoked deep naturalism or resorted to nature as the true (and only) mirror of deep time. Therefore, I adopt two parallel approaches to deep history. One is to see the making of deep time as a reflection of the present in the hybrid assortment of rocks, ruins, fossils, and humans that explorers saw in the nineteenth century. The other is to understand the complex interplay of various disciplines in the nineteenth century.

To give one example, which I explore in detail in chapter 4, the paleontological discoveries of human remains in the 1830s connected the question of human origins to nineteenth-century concerns of tribal aboriginality. Tournal and Jacques Boucher de Perthes proposed the deep geological antiquity of humanity in the 1830s and 1840s, respectively, based on paleontological discoveries of human remains and stone tools. These geological and archaeological explorations of human origin took place at around the same time that European ethnologists encountered different aboriginal populations in Asia, Africa, the Pacific Islands, and South America. These investigations developed a global imagery of human evolution, which, as Chris Manias notes, placed Europeans at an evolutionary advantage to the "savage" tribes.[56] The parallel investigations of aboriginal populations and prehistoric humans suggested that the former were the remains of the latter. More recently, sociological models have been fused with DNA research to form ideas of early human society, kinship, and cultural habits.[57] The chapter locates the problem at the intersection of the disciplines of the deep past such as paleontology, anthropology, and evolutionary biology, which in turn led to the naturalization of erstwhile discussions of human antiquity.

In these instances, the coevolution of anthropology, paleontology, and geology is relatively straightforward. There are several other examples of their mixed methodologies and metaphors. Saul Dubow studied the interplay of the fragments of earth history, anthropology, and material culture in shaping the colonial knowledge of the land and the people of South Africa.[58] The Australian anthropologist Alfred William Howitt practiced what Ian Keen describes as "geologically inspired anthropology."[59] Howitt began his career as a prospector and mining warden in the gold mines in Victoria, Australia. There he met the Aboriginal people and started his research on Aboriginal cultures. Out in the field, he prospected not just for gold but also "mine[d] the Aborigines" in order to record their culture in the same way that he studied fossils—as traces of a lost world.[60] Adolphe Pictet's "linguistic paleontology" treated cultures as fossilized remains and attempted a linguistic reconstruction of the past through geological modes of sifting of cultural layers in search of fossilized remains of extinct languages.[61] In America, where there was no classical archaeology to study, the discipline, in exploring pre-Columbian archaeology, remained closely allied to anthropology. Gordon R. Willey and Philip Phillips (often regarded as the founding figures of New World archaeology) emphatically declared, "American archaeology is anthropology or it is nothing," suggesting that they shared the same goal: the study of prehistoric human evolution.[62]

To give a different example of the coevolution of geological and ethnological categories, "Kemet," the pre-Arabic and pre-Hellenistic name for Egypt, meaning "the black land," has dual connotations. "Kemet" is the term used for the volcanic black soil carried by the Nile to its sprawling plains, vital for growing cotton.[63] Ethnologically, it came to denote the land of black people and in the twentieth century, "Kemet" represented cultural assertions of Afrocentrism.[64] In Gondwana, in the black soil locally called *regur*, which was equally important for cotton cultivation and explorations of human origin, we find similar concurrences. The coevolution of these anthropological and geographical categories established the primeval link between land and people, particularly those who were imagined to be aboriginal to the soil. This assimilation also allowed the colonial states to treat both (the land and its inhabitants) as its subjects.

In several of these instances, nature forms a common point of reference. Seeing the coevolution of these categories through the prism of nature further explains why histories of indigenous populations have been seen to be synonymous with their habitats. James C. Scott has argued that certain communities lived in an anarchist, non-state space, which he calls "Zomia." Zomia carries traces of colonial ethnological and missionary imaginations of the "natural habitat" of tribes. In their paternalistic attempts to adopt and civilize them, the missionaries believed that tribes living in the highlands of central, western, or northeastern India took refuge there from the oppressive Hindu caste system or fled from the Mughal and other precolonial states. The key here is in Scott's use of the term "Zomia." This is not a historical category, nor was it used locally by these people. Rather, it is a modern geographical term coined by Willem van Schendel in 2002 to refer to the huge mass of mainland Southeast Asia that that lies on the margins of several nation-states.[65] Zomia is a metaphorical region invented by scholars who imagined that these people rejected the state. Indeed, the enactment of such a remote and fantastical landscape was fundamental to the thesis of its inhabitants' anarchy. Once again, it is evident how the preconception of geography acts as a determinant of the political history of the region.[66]

This is not to suggest that categories such as the "tribe" or the "aboriginal" were entirely colonial constructions. The concerns here are more with understanding how these categories were used alongside geological ones in colonial literature and how their histories were determined in terms of theories of land formation, which in turn defined their aboriginality.

This narrative of concurrence nonetheless engenders a history of disconnect. In central Australia "Aborigines" were so called to establish their link with the soil at the same time they were being displaced from it, as gold and other minerals were discovered and cattle ranches were established.[67] A similar fate fell upon the tribes of central India who were "discovered" by colonial ethnologists and subsequently adopted by Indian nationalists and renamed Adivasis (a Sanskritic Hindi derivative of the term "aboriginal") in the twentieth century while their habitats were turned into the natural resources, first of the colonial and then the Indian state.[68]

Returning to the question of mixed metaphors, the existing literature has usually followed the histories of geology, anthropology, and archaeology through their respective disciplinary trajectories. There have been some interdisciplinary approaches, particularly in the study of human evolution, through biological, geological, and ethnological investigations, primarily through the collaboration of historians and anthropologists.[69] They generally conform to rather than interrogate the naturalized study of human antiquity.

Similarly, the problems of aboriginality, indigeneity, and primitivism, which are central themes in this book, have been traced mainly within ethnological or anthropological traditions, at least in India.[70] I argue that aboriginality and primitivism were framed at the confluence of multiple disciplines: anthropology, geology, and archaeology. Therefore, it is impossible to appreciate the overlapping narratives of tribes, terrains, and antiquity without understanding the epistemic consiliences at play here.

This is not to suggest that these intellectual crossovers have not been recognized; they indeed have been. However, scholars have not seen them as the key problem. In his innovative work on Indian ethnology, Sumit Guha has used the caste-tribe hierarchy as a metaphor for stratified human evolution in India as seen by British and then by Indian ethnologists, hinting at the geological metaphors implicit in Indian ethnology.[71] He shows how ethnologists identified the aboriginal in the lowest strata of humanity in India. However, Guha leaves the tantalizing possibilities of disciplinary consilience unexplored.[72] Rather than examining the reasons for and consequences of disciplinary convergences, he examines the hypothesis that the tribes were indeed primitive humans.[73]

Manias recognizes the problem in writing distinct histories of the sciences of the deep past.[74] He shows that the use of anthropological and philological methods, along with the discovery of human remains by geologists, shaped the history of human antiquity in Europe. I appreciate Manias's intervention. However, I also suggest that we need to extend the analysis beyond the question of human and racial antiquity. It is important to study the methodological consiliences simultaneously in natural and historical investigations, in convergences of histories of soil, rivers, mountains, and deserts with human histories. In other words, the deep naturalism of the nineteenth century needs to be located simultaneously in its cultural, social, and natural landscapes.

What then, is the disciplinary moment that we find ourselves in? Disciplinarity is a complex historical problem. In his reflections on the subject, Charles Rosenberg shows the complex "ecology" of scientific disciplines, which comprises departments, scholars, the state, and social contexts. Rosenberg has argued that while disciplines shape a scholar's vocational identity, departments provide a tangible "day-to-day" entity to that discipline, making a case for understanding each discipline within its specific contexts.[75] Timothy Lenoir meanwhile has used the "networklike" discursive power of Michel Foucault's biopolitics to highlight the heterogeneity of hospital medicine, statistics, hygiene, and anatomy in the emergence of modern biomedicine.[76] Both Rosenberg and Lenoir draw their conclusions primarily from microbiology, which at the turn of the twentieth century borrowed from diverse fields, including lab-based biochemistry and field-based zoology and parasitology, while serving the interests of the state at the same time. A similar ecology of disciplines can be seen in other sciences such as geology, anthropology, and archaeology in the nineteenth century.

The key focus of the literature on the history of geology in Britain and North America has been its emergence as a "specialised discipline" in the nineteenth century.[77] Histories of geology focus overwhelmingly on the nineteenth century and geology's growing status as a singular discipline. The assumption is that geology emerged as a distinct new discipline in the nineteenth century with the formation of geological organizations such as the Geological Society of London, imperial surveys, publication of specialized journals such as the *Transactions of the Geological Society of London*, appointment of professional geologists, and the proposition of major theories, leaving behind its more eclectic antiquarian engagements of the eighteenth century. This point of view is largely persuasive. However, it can overlook the fact that in practice, geology remained, particularly in the colonies, a hybrid enterprise.

Disciplinary specialization has been analyzed mainly through the three themes of institutionalization, professionalization, and epistemological development. In terms of institutionalization, the Geological Survey of India (GSI) was established in 1851. The Geological Society of London was formed even earlier, in 1807. The British Geological Survey was established in 1835, which subsequently shaped many of the colonial surveys. The Archaeological Survey of India (ASI) was established in 1861, under the leadership of Alexander Cunningham. Ethnology and anthropology have more protracted institutional legacies in India, as they were not formally established as distinct institutions until well into the twentieth century. Anthropology was part of the Zoological Survey of India, which was created out of the Zoological and Anthropological Sections of the Indian Museum in 1916. It was only in 1945 that the Anthropological Section was separated from the Zoological Survey of India.[78] Yet ethnological studies of Indian tribes and castes had started more than a century before. Of these, the GSI had a relatively narrow institutional objective: to identify and extract coal and other mineral resources of India.[79] However, the search for minerals itself was not a limited enterprise. It was laden with ethnological and paleontological experiences, which sustained complex interdisciplinary investigations within the mining enterprises.

Another related disciplinary theme is professionalization, which refers to the emergence of professional geologists, anthropologists, and archaeologists in the nineteenth century. Who were the geologists in the nineteenth century? Who carried out a majority of the geological investigations in remote parts of the world in this period? Were they all geologists by training, profession, and inclination? In the colonies, European explorers often performed multiple roles simultaneously: Orientalist, engineer, ethnologist, missionary, revenue administrator, and mineral prospector. Neither Proby Cautley nor Falconer nor H. W. Voysey, who established some of the foundational aspects of Indian geology in the nineteenth century, were professional geologists, and their work reflected the influence of multiple interests. Others, such as Stephen Hislop, Valentine Ball, and John Henry Rivett-Carnac, whom I engage with in different parts of the book, had diverse intellectual interests and undertook anthropological, geological, zoological, and archaeological investigations simultaneously.

There is a caveat here. The different publications of the GSI or the ASI do not themselves always reflect the eclecticism of these pursuits. Institutional publications such the *Memoirs* and the *Records of the GSI* or the *Asiatick Researches* were often driven by their respective institutional priorities. For example, the revenue department files in the Asia Pacific and Africa Collections (British Library) on the Doab Canal do not contain any information on the fossils discovered during the excavations. These were included in the publications of the Asiatic Society such as the *Journal of the Asiatic Society of Bengal* or *Gleanings in Science*. Similarly, geological publications on the Gondwana strata and fossils do not contain ethnological notes, which are instead included in ethnological publications, often by the same individuals. Therefore, the need is not just to read between the lines, which historians are adept at, but also simultaneously between texts and archives to uncover the tapestry of the new order of things.

Such an approach also reveals the indeterminate epistemological moments. There were several journals and institutions dedicated to geology in Europe, North America, and the colonies by the middle of the nineteenth century. Yet if one reads between texts and archives, it becomes evident that nineteenth-century geological theories such as the existence of Gondwanaland, land bridges, Lemuria, or glaciation and contraction were shaped by contemporary ethnological, zoological, and botanical theories and hence by holistic observations of the earth across different lines of inquiry. It can be argued that it was only with the acceptance of the continental drift theory in the 1950s—which suggested that massive continental plates had moved, an idea almost inconceivable in the nineteenth century, as I discuss in chapter 5—that geology came into its own as a discipline. It was only then that it acquired the planetary vision that we associate with it today, through which its previous earthly encounters with other disciplines as well as earthly inhabitants appear insignificant. As the British geologist Tony Hallam suggested, the drift theory "changed the very fabric of Earth science."[80]

Ethnological studies of Indians as different "races" began in India in the late eighteenth century from biblical, Orientalist, and Aryan-Dravidian cultural premises.[81] By the late nineteenth century, they came to be structured more around the study of tribal societies. Yet as we shall see in chapters 4 and 5, ideas of tribal aboriginality remained deeply infused with those of the geological primitivism of the landscapes that they lived in, as well as biblical ideas of Hamite migrations from Africa to Asia, which ethnologists and missionaries believed peopled the southern continents.

Then there is the problem of prehistoric archaeology. What came to be seen as prehistoric archaeology in the twentieth century, that is, the study of prehistoric relics and stone implements, was in the nineteenth century firmly the domain of those who worked primarily on geology, archaeology, and paleontology, such as Lyell, Joseph Prestwich, Falconer, Robert Bruce Foote, and Boucher de Perthes.[82] Rudwick notes these disciplinary crossovers in his discussions on the "archaeology of the earth," where he refers to Johann Blumenbach's use of the term *archaeologia* (which in the eighteenth century meant antiquity in general) for his study of fossils.[83] In doing so, he shows that Blumenbach established fossils as objects of general antiquarian studies in the eighteenth century. However, he does not comment on the significance of the thematic crossovers and convergences that continued in the nineteenth century, when both archaeology and paleontology had seemingly adopted distinct antiquarian pursuits.

In India, archaeology as a discipline and the ASI as an institution focused more on classical antiquity in the nineteenth century. This was evident in the works of its main patrons, James Prinsep, Wilson, and Cunningham, who studied Buddhist and Asokan inscriptions, numismatics, and architectural remains and compared them with Roman and Greek ones.[84] Yet, as evidence of the contaminated experiences of the past, the very same sites also revealed prehistoric fossils and urban settlements. The Indian archaeologist Rakhaldas Banerjee came across the seals of the Mohenjo-daro belonging to the Indus Valley civilization while digging for a Buddhist stupa in Sindh in northwest India in 1921–22.[85] Adjacent to the ruins of the highly evolved and urbanized Indus Valley civilization, geologists found rock art belonging presumably to prehistoric humans.[86] However, most of the discoveries of stone tools in the nineteenth century in India by Foote, Ball, and Rivett-Carnac fell simultaneously within geological and ethnological research as they compared them with the fossils they were found alongside and with those used by contemporary tribes. Even in the early twentieth century, Foote's entire collection of prehistoric stone implements was preserved and displayed in the Art and Ethnological Sections of the Madras Museum.[87] Taken together, these examples illustrate that the nineteenth century, besides being the period of professionalization and specialization, was the period when the greatest synthesis of ideas between these disciplines took place. I trace how in this interaction, nature emerged as the common thread in the study of the past.

Because of the concurrent nature of these histories, there is no single entry point to the narratives of this book. At least three distinct passages lead to the history of deep naturalism of India: the Himalayan expeditions across Tibet and Kashmir and the earliest discoveries of fossils by British explorers; the eighteenth-century Mosaic historical traditions in India, through which geologists linked Indian fossils with Puranic traditions; and the geological and ethnological explorations of landscapes and tribes in central India. Instead, this book begins with the story of the canal of Zabita Khan, the construction of the Yamuna Canal, and the making of colonial archaeology in the Doab region. This is to demonstrate how the deep past of nature came to define and dominate our understanding of the past *in general*, even the past of nonnatural or non-geological objects (such as that of the canal).

CHAPTER ONE

The Canal of Zabita Khan

The Nature of History

The deep past of India was founded on the medieval remains of the Delhi Sultanate and Mughal Empire. This deep past is disconcertingly new, characteristically obliterative, and yet uncomfortably enduring. It also lies atop ruins. The story of how deep time superseded various other historical imaginations in India structures my critique of the conceptions of deep history, of the ways they overwrite other histories. By tracing this process through the history of the canal of Zabita Khan, this chapter analyzes how a colonial engineering project became an archaeological and a geological one and enabled the construction of the prehistoric antiquity of the Indian subcontinent.

In the early nineteenth century, the British began one of their most ambitious irrigation projects in India. This was in the Doab (Do-ab: literally "two waters," the tract between the Rivers Yamuna and Ganga) region of northern India (fig. 1.1). The East India Company (EIC) occupied Delhi on the banks of the Yamuna as well as the entire Doab and Delhi-Agra region in 1803 during the Second Anglo-Maratha War (1803–5). The company's interest in the canal was initiated soon after surgeon Graeme Mercer (1764–1841) offered to reopen the old Delhi canal at his own expense on the condition that he would enjoy the entire profits from its commercial use for the next 20 years.[1] Mercer had participated in the Second Anglo-Maratha War and entered Delhi with the British troops. The government rejected his suggestion. In 1808, the chief engineer of the EIC at Agra, General Alexander Kyd (1754–1826) surveyed the entire Doab region, following the course of the Yamuna from the Himalayan foothills into the plains.[2] As other British officers surveyed the land as well, they found traces of an ancient canal system throughout the region. Kyd explored the possibility of using the old bed of the canal for the new one. He believed that much of the upper part, around the town of Saharanpur in the Himalayan foothills, was in good condition and could be used for the new canal.[3]

The existence of the old canal network highlighted the need for the new one. The Board of Commissioners of the EIC wrote to the governor in Calcutta that the land, at present desolate and barren, bore traces of a fertile past.

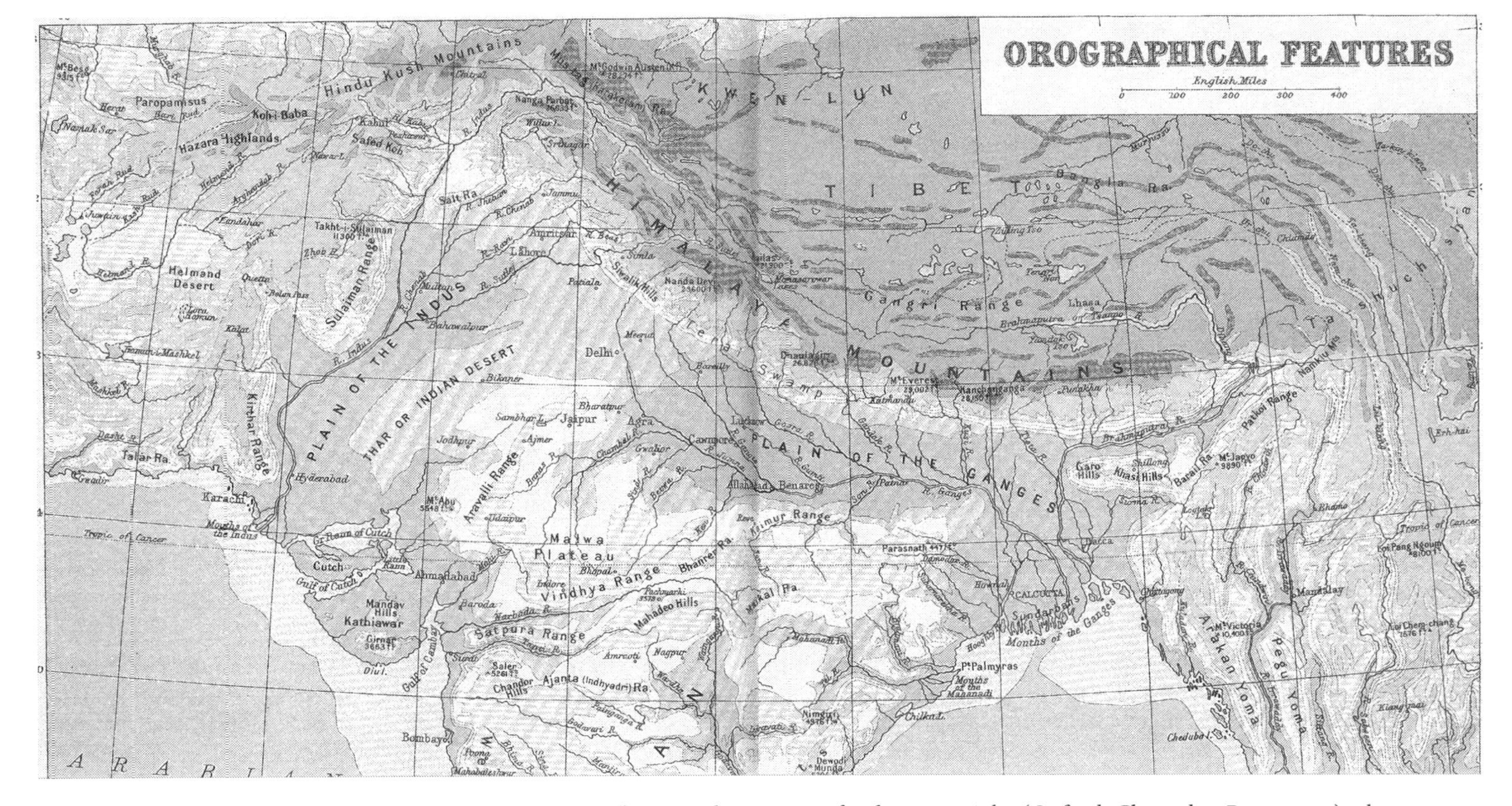

Fig. 1.1. The Ganga Yamuna Doab. "Orographical Features," *Imperial Gazetteer of India*, v. 26, *Atlas* (Oxford: Clarendon Press, 1909), plate 4.

In these deserted tracts "groves of mango trees" served as evidence of the former prosperity of the region, and "even where no vestige of human industry remains, the names of the villages which once stood there are still preserved in the public accounts."[4] The decaying landscape reflected a lost past in need of regeneration. A new canal network, the board stressed, would revitalize the landscape.

As British engineers began to dig the canal, they found traces of a medieval network of canals around the entire Doab region. In 1809, the Board of Commissioners asked Lieutenant James Tod (whom Mercer had familiarized with the geography of the Doab and who later became famous for his work on the antiquity of Rajasthan) to survey the Doab Canal. Tod noted that "very old men" who lived in the area, which was once prosperous and ruled by the Mughals and the great landholders, spoke about the former canal.[5] According to him, the Mughal emperor Muhammad Shah (1702–48) built a small canal and a dam near Faizabad (a small village 26 miles north of Saharanpur) in the Himalayan foothills to channelize the water from the river to the canal.[6] Subsequent rulers of the region repaired and restored the canal system. The British referred to this as "the canal of Zabeta Khan," because the Rohilla chief Zabita Khan (who died in 1785) was the last ruler known to have restored it, following the practice of his father, Najib-ud-Dawlah.[7] The Rohilla chiefs controlled parts of Delhi and the Upper Doab before the Marathas in the late eighteenth century. The Marathas, who originated from western India, controlled major parts of central, western, and northern India, following the fall of the Mughal Empire.

The colonial engineering project was simultaneously archaeological, historical, and geological. The British discovery of Indian antiquity, or the genesis of the modern historical imagination of India, is seen to lie in the Orientalist textual traditions of William Jones and others in the late eighteenth century, traditions that were gradually reinforced by archaeological and architectural explorations.[8] This journey into Indian antiquity was, however, more complex. The construction of the colonial canal also became a restoration project of the ancient one. This, in turn, led to colonial archaeological and geological projects around the Yamuna and in the Himalayan foothills. As colonial engineers excavated the old canal network, they explored textual references to them, thereby returning the archaeological search to the textual. Around the same time, British scholars were translating significant texts of the medieval period such as *Tarikh-e-Farishta* and *Chachnama*, which provided historical context to the architectural remains. Digging for the ancient canal also unearthed prehistoric fossils and cities, leading to formative ideas on Indian antiquity, prehistoric geography, and human evolution in the subcontinent. There were two processes at play here. First, at the level of practice, building the canal became an archaeological and a geological activity. Second, at the level of epistemology, discursive connections were made between monuments and geographies and the textual traditions of Indian antiquity.

The canal is also one of the sites where antiquity became naturalized as deep history. On the banks of the Yamuna, the canals, rivers, and natural and historical antiquity became subjects of the *same* ecumenical inquiry. Over the course of the excavations, ancient canals appeared indistinguishable from rivers. The discovery of fossils, possibly of humans, in the Yamuna riverbeds, aligned human antiquity with the evolution of the terrain. In all these accounts, the lines between the monument and the land, or the "historical" and the "natural," were imperceptible. Rivers moved; legends moved with them. Dead riverbeds became canals; canals became natural channels of water; fossils embedded in their sands invoked deep Hindu mythologies of the land (as we will see in chapter 3). In the process, the landscape, the legends, the fossils, and the monuments were all aligned with deep colonial antiquarianism. Although it sounds cumbrous, "natural-historical" is the only phrase that can explain this phenomenon. It signifies the convergence of natural and historical imaginations in the study of antiquity simultaneously in wandering rivers, dead riverbeds, canals, myths, folklores, and texts. The phrase allows us to understand the composite form of antiquarianism situated beyond geology or natural history and yet inherently shaped by them. As this chapter will show, the naturalization of the canal took place through this endeavor to locate the past in composite sites. Every other term, such as "geological," "natural history," "biological," "geohistory," and even "history" has acquired a specific meaning, mainly due to the tyranny of academic and institutional disciplines. The nineteenth century was, to repeat, not just the moment of the formation of disciplines; it was also the period of sharing the ways of knowing the past.

The region around the Yamuna was the site of the concurrence of medieval, ancient, and deep histories of India. Delhi and several major historical sites of precolonial India, such as the Taj Mahal, Vrindavan, and Qutb Minar, are located along the banks of the Yamuna and in the Doab region. The Doab was the traditional seat of political power in the fertile Gangetic river system of northern India. Between the Sutlej, the easternmost river of the Punjab, and the Yamuna, the westernmost river of the Gangetic Plain, lie the vast arid and semiarid regions of western and northwestern India, where repeated migrations into and invasions of the northern Indian plains occurred. Therefore, this intersectional region is associated with debates on the supposed Aryan origins of the Hindus, the Muslim invasions of the subcontinent, and the subsequent Hindu-Muslim confrontations that were reimagined in the nineteenth century. It was also the site of several mythical and historical conflicts, such as those in Kurukshetra, Panipat, and Tarain, which mark the borders of northern Indian cultural and political identities. The discovery of the prehistoric Harappan civilization along the Indus River in the early twentieth century further intensified the debate as to whether the Aryans had indeed invaded India and displaced its indigenous populations. Almost every river in the region, real or imagined, is seen to bear witness to these conflictual histories.

In narrating this history of overlapping geological and historical antiquities on the banks of the Yamuna, I acknowledge the significant scholarship on nineteenth-century antiquarianism. Scholars have identified the eclectic ways philologists, archaeologists, ethnologists, and Indologists have approached the past through texts, languages, numismatic studies, and historical monuments.[9] They have also traced the progressive discovery of deep archaeological antiquarianism in India.[10] Here, I will chart a narrative different from such histories of Indian antiquarianism or of Indian archaeology. Through an exploration of the interplay of "historical" antiquarianism with the "geological" one, I trace how deep history imposed itself on other histories, through the metamorphosis of the canal of Zabita Khan into the perennial river of India with a Hindu identity by obliterating its plural pasts.

The Canals of Paradise: Mapping Landscapes, Mapping History

Tod completed his survey in 1810. However, the plans were shelved, and it was only in 1822 that another engineer, Lieutenant Henry De Bude of the Bengal Engineers, along with Robert Smith as the superintendent, was appointed to survey the practicality of opening the Doab Canal.[11] De Bude left India soon, and in 1823 Smith succeeded him as the superintendent of the Yamuna Canal project. Smith made plans for digging the canal from Faizabad, the site of the original cut of the old canal from the river.[12] Under his supervision, the work became, as a colonial engineer described, "practically a restoration of the old Mogul canal."[13]

The mapping of the Yamuna Canal was not a mere topographical exercise; it was also a mapping of its antiquity. This was a form of "deep" mapping, through the excavation of the landscape itself. Robert Smith was an engineer by training. He was also a skilled painter and one of the several British artists and draftsmen in British India, such as Thomas Daniell and George Chinnery, who vividly painted the rapidly changing historical landscapes of the northern plains in the early nineteenth century. Smith received instruction in landscape painting early in his career from Chinnery in Calcutta.[14] As he traveled through northern India, he sketched the sacred banks of the Ganga in Hardwar, Aurangzeb's Mosque at Varanasi, scenes of Allahabad, and of course the Yamuna Canal. Around 1815 he painted a brooding image of an old crumbling tomb in eastern India from which a large banyan tree had grown. This painting captured both the state of the decaying medieval empires that the British found in India and the metaphoric representation of the epistemic incongruity that British engineers, Orientalists, and scientists faced: the intertwining of encroaching nature and the degenerating monument, the antiquity of one inseparable from that of the other. When he moved to Delhi and began work on the old Doab canal network, Smith became increasingly interested in restoring the dilapidated Indo-Islamic monuments in and around Delhi, such as the Qutb Minar and the Jama Masjid.[15]

The restorations also revealed and redrafted the contested and overlapping histories of these monuments. British engineers first wrote about the Qutb Minar in the late eighteenth century, when James T. Blunt, an engineer, provided an account of its dimensions.[16] It was only in the 1820s, when Smith began to rehabilitate the structure, that the British seriously investigated its history. Walter Ewer studied the various inscriptions on the minar and other ruins in its vicinity with the help of a telescope.[17] The inscriptions revealed that the minar was built during the rule of Sultan Iltutmish in the early thirteenth century and was subsequently repaired by a later ruler of Delhi, Sikandar Lodhi in 1503.[18] Ewer also noticed the iron pillar near the minar, which was surrounded by two arcades of pillars that resembled Hindu architecture. This juxtaposition of Islamic and Hindu structures intrigued him, and he explored the possibility that the minar was built at the site of a Hindu temple. The Hindu inscriptions around the pillar suggested that there was indeed a temple at the site. However, Ewer dismissed the suggestions made by his Hindu informers that the Qutb Minar itself was built by a Hindu prince and later decorated by the Turkish sultans of Delhi.[19] The materials used in the Qutb were consistent with other structures built by the Turkish rulers around that time. He also noted that the minar was in desperate need of repair; major cracks where a banyan tree had taken root threatened the entire structure. In 1826, Smith made substantial repairs to it as well as the Jama Masjid, the main mosque of Delhi.[20] As he restored the minar, he added a British dimension to it: a cupola at the top of the minar to replace the damaged structure. This was deemed unsuitable and was removed in 1848.[21]

The imagination of the past in this rapidly transforming landscape fused the historical with the imaginary. Smith's painting of the Yamuna Canal (fig. 1.2) was in fact a work of fiction. It is dated 1808, when British engineers had started the exploratory works and before he even joined the project.[22] Yet Smith painted the canal with sailing boats. From all available historical records, it is apparent that the old canal, as Smith and other engineers found it, was not navigable by any vessel at that time.[23] By the time water was finally released into the canal more than two decades later in 1830, under Proby Cautley's stewardship, Smith had taken sick leave and returned to England.[24] This futuristic depiction of the ancient canal, central to the British journey into Indian antiquity as well as to the future of the colonial agrarian economy, demonstrates the complexity of the British imagination of India, imbricated with notions of the past, the present, and the future. In the work of several other colonial engineers involved in the canal project, as we shall see, the restorations of ancient monuments were juxtaposed to analyses of future land

Fig. 1.2. The Jumna Canal near Meerut with soldiers and fortifications, by Lieutenant Robert Smith, 1808. Oil on paper. Courtesy of the Council of the National Army Museum, London.

reforms and revenue systems. The British saw themselves simultaneously transforming and restoring the landscape. Throughout the British discovery of these ancient canals, the historical coexisted with the imagined.

As the British reading of the history of the Indo-Gangetic Plain went deeper, so did the antiquity of the canal. The canal gradually became one of the conduits of the British discovery of Indo-Islamic antiquity in the Doab region. In 1825 another engineer, Cautley, joined the project as assistant to Robert Smith. He realized that this was not an eighteenth-century canal, as previously assumed. Zabita Khan had merely restored it. Cautley was initially uncertain as to who had originally built the canals, the Mughal emperor Shah Jahan in the seventeenth century or Muhammad Shah in the early eighteenth century.[25] While digging the canal he began to read John Briggs's recently published *History of the Rise of the Mahomedan Power in India* (1829).[26] The work was translated from Mahomed Qasim Farishta's late sixteenth-century text *Tarikh-e-Farishta*, originally written in Persian. Cautley became aware that the British were working on a project that seemed to have been undertaken in the fourteenth century by the Sultan Feroze Shah Tughlaq (1309–88) of the Tughlaq dynasty of Delhi. Feroze Shah Tughlaq began the project of irrigating the arid region between Delhi and Punjab mainly through canals drawn from the Yamuna. As he dug a canal from the Yamuna, he established the city of Hisar (in Haryana) west of Delhi, then known as Hissar Ferozah.[27] Subsequent rulers of Delhi maintained and dug the canals until the eighteenth century. In his *Memoir of a Map of Hindoostan* (1788), James Rennell too referred to Feroze's canals.[28] The information, however, was derived from textual sources, and the canal remained confined to cartographic descriptions. Cautley's was the first occasion on which British engineers physically excavated the ancient canals.

British engineers also realized that there were not just one but several canals emanating from the Yamuna. They were known by different names, often after the rulers who built or restored them in different periods: Zabita Khan's canal (the Doab Canal, between Ganga and Yamuna), Feroze Shah's canal (the one flowing westward from the Yamuna into the semiarid region), and Ali Mardan Khan's canal (that supplied water to Delhi, also known as the Delhi canal).[29] Of these, British engineers found the Delhi canal in the best condition, as it was built "with a most lavished hand," was maintained by subsequent rulers, and could be reopened relatively easily.[30] By the end of May 1820, British engineers succeeded in passing water through the Delhi canal. The restoration work on Feroze Shah's canal began in 1823 under Robert Smith's supervision, and in 1825, water was channeled through the canal. The restoration of the Doab Canal, which ran along the eastern side of the Yamuna starting from the old cut at Faizabad and rejoining the river at Delhi, began in 1824 and was completed in 1830.[31]

The intricate canal system around Delhi built by Feroze Shah Tughlaq and restored by the Mughals or their successors was known as the Nahr-i-Behist, or the Canals of Paradise. They served both aesthetic and utilitarian purposes, providing water through beautifully ornate channels in the imperial city as well as through long stretches of waterways to the surrounding and predominantly arid region. The Mughal emperor Shah Jahan's Delhi was one of the most beautiful cities in the world in the seventeenth century. The Delhi canal, originally built by the Tughlaq dynasty between the fourteenth and fifteenth centuries and renovated by the Mughal administrator Ali Mardan Khan in the early seventeenth century, ran through the center of the main street of Shah Jahan's capital. It was lined with stones and opened into square or octagonal pools at regular intervals. At night, moonlight glistened like silver on the water, so the main street was called Chandni (silver/moonlit) Chowk. The canal also carried water to large houses in Delhi through underground channels.[32] The British reopened parts of the Delhi canal but closed it permanently in the middle of the nineteenth century because of health concerns. Excavating the ruins of the Canals of Paradise created a route for the British to journey into Indo-Islamic history and, eventually, into the deep past of India.

The Archaeology of Ruins

Delhi was a relatively insignificant city in the early nineteenth century. The centers of power had long shifted to Lucknow, Pune, Mysore, and the colonial port cities of Calcutta, Bombay, and Madras. The Delhi that the British occupied in the early nineteenth century was a city of ruins, caught between two empires, the Mughal and the British, and between the old and the new.[33] Colonial engineers were enchanted by this old and decaying Delhi. Their appreciation of its precolonial aesthetics, architecture, and engineering works was a Romantic one. They also felt a sense of regret for it as a lost world whose traces were rapidly disappearing because of their own engineering and urban projects.[34] Almost every engineer who worked on the Yamuna Canal—Smith, Cautley, John Colvin, and J. A. Hodgson—took a deep interest in the archaeological remains of Delhi and the Doab region. The British would later ransack the city and destroy most of its precolonial architecture in retaliation for the revolt of 1857. However, the early nineteenth century was still the age of Orientalist Romanticism.

There were two different imageries of ruins at play here. Cities, ruins, and conquests featured significantly in the cultural imagery of northern India in the nineteenth century. To the Urdu- and Persian-speaking literati of Delhi, ruins embodied the cultural and political emptiness of the landscape as the British became its new rulers. The celebrated nineteenth-century Urdu poet Mirza Ghalib, who lived through the transformation of Delhi, lamented, "Not everything has become manifest in the tulip and the rose. What forms must there be hidden in the dust!"[35]

To the British, meanwhile, the ruins around the Yamuna held promises of a Romantic and classical past. It was on this desolate landscape of mango groves, ancient cities, and dead canals that the discipline of colonial archaeology built some of its foundations. The Archaeological Society of Delhi was established in 1847, almost two decades before the establishment of the Archaeological Survey of India, as a culmination of these discoveries of Indo-Islamic heritage along the banks of the Yamuna that had begun with the excavation of the Doab Canal.[36] The same year, the eminent scholar Syed (later Sir Syed) Ahmed Khan published *Asar-al-Sanadid* (History of monuments of Delhi).[37] *Asar* represented the nineteenth-century genre in Urdu literature of ruins as desolate and bereft. Khan immortalized the city of Shahjahanabad in Delhi built by the Mughal emperor Shah Jahan in the seventeenth century. C. M. Naim notes that in the later edition of *Asar*, published in 1854, *after* the formation of the archaeological society, Khan adopted a more distinctly archaeological language.[38] The ruins were now represented as Romantic and antiquarian, not just as sites of lament and desolation.

This Romanticism was shaped by the contemporary British discovery of the Indo-Greek heritage through archaeological remains in northern and northwestern India from the 1830s, particularly in the works of James Prinsep and H. H. Wilson.[39] The archaeological remains of the Doab suggested the presence of older cities and urban civilizations beneath the medieval ones. In March 1835, Tod, now retired, wrote from the Piazza Barberini in Rome, reflecting on his earlier explorations around Delhi. He suggested that Alexander Burnes and James Gerard, who at that time were tracing remains of Indo-Greek heritage in the vast region from Afghanistan to Bukhara, could find similar remnants of the classical past in the ruins around the Yamuna as well:[40]

> Let not the antiquary forget the old cities on the east and west of the Jamna, in the desert, and in the Panjab, of which I have given lists, where his toil will be richly rewarded. . . . A topographical map, with explanations of ancient Delhi,

> is yet a desideratum, and of the first interest: this I had nearly accomplished during the four months I resided amidst the tombs of the city.[41]

In this landscape of ruins, Tod had discovered ancient cities along the Yamuna where he found coins signifying traces of Greek conquest of the region.[42] British archaeology found a deeper classical heritage in the Doab.

Since colonial archaeology emerged as part of the establishment of the new empire on the ruins of the old, antiquity there was juxtaposed to the modern throughout the history of the Doab Canal. The restored canals, irrigation channels, and bridges engraved a story of loss and regeneration into the landscape. Cautley also refurbished Mughal monuments around Delhi while he repaired the old canal.[43] As he began work on the Doab Canal in 1826, he visited the dilapidated Badshah Mahal across the Yamuna on the other side of Delhi.[44] The mahal was the hunting palace of the Mughal rulers. One of the original cuts for the Doab Canal was made there and supplied water to the mahal. Cautley inspected the aqueduct of the palace to determine the direction of the watercourse of the ancient canal.[45] He found the palace covered in an "impenetrable jungle." The ornamental stonework had been removed by the locals. After clearing the dense vegetation, he discovered ornate marble floors that had escaped plunder because of the thickness of the debris and overgrowth. He removed the largest marble reservoir to decorate the EIC's new botanical garden being built at that time in Saharanpur. He also appropriated the smallest reservoir for the fountain on the Rajpoor watercourse at Dehra Dun, the plans for which he drew up as part of his Yamuna Canal project.[46] The masonry of the old canal ornamented the modern one.

The Doab Canal enables us to recognize some of the historical processes through which Indian antiquity was shaped by the constant juxtaposition of the ancient and the modern. Even Arthur Cotton's famous Kallanai Dam, built circa 1830 across the Kaveri River in the Madras Presidency, otherwise seen as the most intrusive colonial intervention on the landscape, is also viewed as a renovation of an ancient dam almost 2,000 years old.[47] The Chola kings were supposedly the original builders of the dam, known as the "Grand Anicut."[48]

This overlapping of the past and the present leads to the question, how should we understand the history of these canals? One possibility, following the suggestions of David Arnold, is to see colonial canals as a critical historical juncture of colonialism "transforming" the Indian landscape.[49] Other historians have similarly seen colonial irrigation projects as critical "transformative technologies" that altered the landscape, the courses and drainage of rivers, and eventually agricultural patterns.[50] It is also possible to see the simultaneous archaeological interest as colonial Romanticism, against the motif of improvement that the British were simultaneously pursuing in India.[51]

At the same time, it is tempting to see these restored canals in terms of continuity between precolonial and early colonial projects, knowledge systems, traditions, and institutions, following the suggestions of C. A. Bayly.[52] Richard Grove likewise emphasized this continuity when he suggested that the Saharanpur and other EIC gardens were established on the foundation of older Mughal gardens and followed precolonial architecture and aesthetics of nature.[53]

Although elements of both these analyses are useful, neither fully explains the history of the Doab Canal. On the one hand, there is a history of continuity there, as digging for the new canal was part of the restoration of the old one; British rulers maintained and created new cuts of the old canal, just as their precolonial predecessors had. On the other, there is a significant disjuncture ushered in by colonialism and the new disciplines that it introduced, such as history, Indology, and archaeology, which invested new meanings in the old riverbeds, canals, and ruins. This was indeed a transformation, but not, as Arnold suggests, through the binary of improvement and nostalgia. There was no such binary; the two were symbiotic. At the same time, the continuity suggested by Bayly and Grove appears superficial, not least because it is difficult to imagine continuity in a history situated on ruins. The precolonial restorations of the canals or cartographic exercises by Mughal and Sultanate rulers were not associated with parallel references to texts or archaeological excavations. In other words, they had remained chiefly engineering and cartographic projects, not marked by antiquarianism.

There was something more significant taking place here than "continuity" or "change." It was the British discovery of Indian antiquity, which created a new order of things in which the past, present, and future were fused inseparably. Here I need to correct a point I made in my earlier work, where I suggested that in the nineteenth century, the antiquarian projects of the Asiatic Society became distinct from its scientific explorations in botany, geology, and meteorology (which by the middle of the nineteenth century moved to specialized institutions or societies), rendering the society an exclusively antiquarian institution.[54] Instead, here I argue that Indological antiquarianism was deeply infused with the scientific and engineering projects of the nineteenth century, and vice versa. The British were not simply restoring canals or drawing maps. They were also investigating the antiquity of the Indian landscape. Their colonial engineering and survey projects were therefore invested with both antiquarianism *and* scientific modernity.

The excavation for the new canal or the restoration of the old was not an exclusively or typically archaeological or geological excavation—a progressive journey through layers of the past and a gradual deepening of time. British engineers simultaneously built a canal for the benefit of the colonial economy, restored ancient monuments in the Doab, and, as we shall see, deliberated on the origin of humans in the region. The process of building the Yamuna and Ganges Canals familiarized the British with the lay of the land, the slopes, the *khadirs* and *bangars*,[55] the ancient irrigation systems, and the antiquities and myths. As Cautley engineered the Doab and the Ganges Canals, he discovered old irrigation channels called the *rajbuhas*.[56] He adopted them as part of his colonial irrigation project, known as the *rajbuha* system, to construct a subsidiary system of minor channels to carry water from the main canal to the agricultural fields.[57] Since the alluvial clay of the Doab was soft, piling the canal banks (as was done in England) was not possible. Cautley and Colvin borrowed the indigenous method, which Cautley described as "the system of hydraulic architecture of Upper India" of sinking wells. This involved the use of masonry walls and underwater divers who extracted the soil.[58] As Cautley wrote, "The Doab Canal works have paid equal homage to this admirable native conception."[59] As colonial officials created the first cut of the Ganges Canal at the Ganesh Ghat in Hardwar, they built steps for the use of pilgrims who bathed in the holy water.[60] While doing so they observed the bathing rituals at this sacred Hindu site.[61] During digging for the Ganges Canal near Kanpur, Cautley found the artificial mounds built of *kankar* (calcium carbonate nodules) along the rivers, which revealed Hindu sculptures and remains of ancient Hindu temples underneath.[62] Between Hardwar and Kanpur, the Ganges Canal also unearthed ancient seals.[63] As he built the Yamuna and Ganges Canals, Cautley established plantations of sal, sheesham, and teak trees along their banks, which came to be vital for the supply of timber to the British Empire.[64] These concurrences of the future, present, past, and deep past provide a clue to the colonial journey into Indian antiquity, innately conditioning it to the shifts between the ancient and the modern.

John Colvin, another engineer who extended the canal project west of Delhi and constructed several bridges and drainage works on it, studied the entire ancient canal network to the west of the Yamuna. His main task was to survey Feroze Shah's fourteenth-century network of canals along the western Yamuna, which was around 240 miles long. He found that in some places the old canals were indistinguishable from ancient riverbeds.[65] According to him, Feroze Shah's canal had in effect become the westernmost channel of Yamuna. He found several other canals that appeared to him to be branches of smaller streams.[66] From Dhatrath (around 100 miles northwest of Delhi), the canal was seemingly dug over the bed of an "ancient river," which was cleared out to convey water to the city of Hisar. From there, the old canal ran toward Bikaner to the west before disappearing in the desert sand. Another canal was dug into the bed of the ancient river to pass the surplus water of the canal a few miles beyond Hisar.[67] That riverbed then ran westward along the northern bounds of the sandy desert until it united with another canal from the Ghaggar River, near Badhopal. Colvin referred to the drying up of the Ghaggar and the depopulation of the region, which, he speculated, was due to the lack of water.[68] Several rivers and canals on Colvin's map (fig. 1.3) were not *present* in the landscape that he studied. He imagined them, basing his speculations on local folklore and historical texts. The River Ghaggar was traced on Colvin's map, although the track of the riverbed was often lost in the sand. The river, as British explorers discovered, was at the same time well entrenched in popular memory.[69]

In drawing up his map of ancient canals and rivers, Colvin followed Robert Smith's mode of conflating the historical with the imagined. On Colvin's map, canals, such as that of Feroze Shah, merged seamlessly into rivers and *rajbuhas*. The Delhi Canal itself is shown to have changed course, just like a river. It was as if the distinction between the historical canals and natural riverbeds became irrelevant in his map. In fact, in the textual traditions of the Delhi Sultanate, from which the British engineers derived their information, canals were considered perfect if they were close to being "natural" because the "naturalness" demonstrated that the monarchs had used their skills and enterprise much like the Divine Creator. Canals were sometimes given names, just as if they were rivers.[70]

In the explorations of the ancient canals of the Yamuna, British engineers made frequent references to mythical rivers, cities, and lost civilizations. Henry Yule, another engineer and Orientalist scholar, spent several years restoring and developing the Mughal irrigation system of northern India while working on the Ganges Canal project.[71] As he surveyed the old canals, Yule came across a decree issued by the Mughal emperor Akbar (1542–1605).[72] The decree was dated 1568 and was issued at Ferozepur.[73] In it, Akbar announced the reopening of parts of the canal initially dug by Feroze Shah. This canal later came to be known as Chitrang Nadi (River Chitrang), which channelized

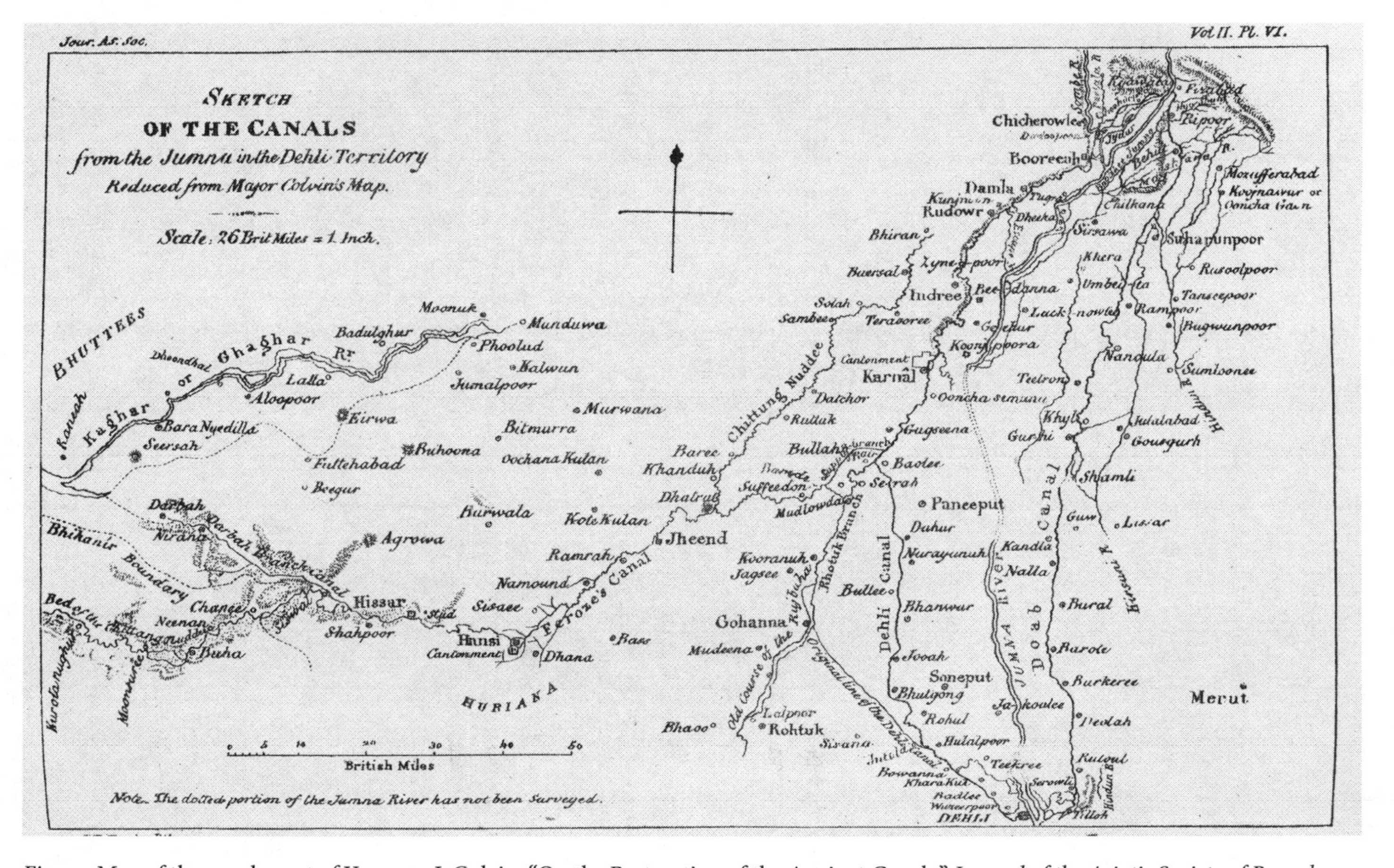

Fig. 1.3. Map of the canals west of Yamuna. J. Colvin, "On the Restoration of the Ancient Canals," *Journal of the Asiatic Society of Bengal* 2 (1833): 105, plate 6. Image copyright of the University of Manchester.

water to the westernmost parts of the semiarid region between Haryana and Rajasthan.[74] Besides diverting enough water into the canal so that boats could ply it, Akbar also ordered new bridges and banks be constructed on the canal. Of these, Yule found that the bridges at Gharaunda and Safidon (both near Karnal in Haryana) had survived, "both massive structures with pointed arches."[75]

As Yule navigated through Akbar's canal decree—which he corroborated with Alexander Dow's translation of Farishta's history, Rennell's *Memoir*, the English translation of Ibn Battuta's *Rihla*—along with the archaeological remains and the complex mesh of canals and rivers, he discovered intricate arrangements of rivulets, canals, dead riverbeds, and folklore of mythological rivers in the western Yamuna region.[76] The city referred to as "Sirsa" appeared to him to be the Sursuttee of the past, through which, according to Farishta, Feroze dug his canal to Hisar around 1354.[77] He identified that canal with the one then known as the Choya Nala (Choya Stream). The name Sirsutti seemed to him to be a derivation of Saraswati, the renowned mythological river, which also appeared to give the town of Sirsa its name. He found that another engineer, Major Brown, had tracked an old channel of the Ghaggar in the direction of Hisar, which the locals referred to as the old bed of the Saraswati. An amused Yule commented, "The Sirsutti has a gift of ubiquity!"[78]

The modern imagination of the Saraswati became possible with the insertion of deep geological thinking into the existing antiquarian imagination and the subsequent confluence of the geological, the mythological, and the historical. The Saraswati, the mythical river of the Vedic texts, which was supposed to have dried up or become "lost," was reimagined in this landscape of loss by nineteenth-century British natural-historical antiquarianism. The history of the river and the debates on its original course became increasingly complicated in subsequent decades, with the addition of layers of geological discoveries of riverbeds, the archaeological findings of prehistoric Harappan sites, and the textual interpretations of the advent of Aryan civilization of India.[79] It is important to situate the discovery within the journey into Indian antiquity as it unfolded in the nineteenth century. The search for the mythical river simultaneously within texts and in the landscape began with the excavations of the old canal network.

The Ubiquitous Saraswati

The mythical and "lost" Saraswati has provoked considerable debate in recent years. It has also generated considerable literature.[80] This chapter does not address the accuracy of the claim that the Vedic Saraswati was the modern

Ghaggar or Hakra. I am more interested in investigating the genealogy of its discovery in the nineteenth century. The modern enactment of the mythical river is situated in the natural-historical journey into India's antiquity on the banks of the Yamuna. The beginning of the debate around the Saraswati is usually located in late nineteenth-century translations of Vedic texts and the attempts at identifying the Vedic river in the modern landscape.[81] Yet the first mention of the mythical river in the modern era was in fact not in Indological literature. Early in the nineteenth century, British writers found traces of the Saraswati (or its derivative "Soorsutty") in two different sites: *Tarikh-e-Farishta* and in the excavations of the Yamuna Canal. They were simultaneously navigating, as we have seen, through landscapes, local folklore, and historical tracts. There the Saraswati, or more precisely "Soorsutty," was not a mythical river but one that was partly dried or buried, like many others in this region west of the Yamuna.

Embedded in this history is the narrative of the establishment of Muslim rule in India, which was being retold in the nineteenth century as a tale of Muslim invasions of a Hindu landscape. The Indologist John Briggs translated Farishta's sixteenth-century history of India in 1829 and renamed it the *History of the Rise of the Mahomedan Power* (the literal translation of the original Persian title would be "Farishta's history"). Briggs narrated the story of the Muslim invasion of India, in which the Soorsutty was a key point of reference. In several places, the book refers to the river and the city, both by the same name. According to Briggs, in 1191 Muhammad Ghori from Ghor (in present-day Afghanistan) fought the decisive battle against Rajput rulers on the banks of that river as his forces invaded India. The respective armies were camped on either side. Ghori forded the Soorsutty with his army and entered the Rajput camp just before dawn. After a ferocious battle, in which several of the Rajput princes were killed, the fort of Soorsutty fell into the hands of Ghori. These clashes are significant episodes of Indian history, usually regarded as the points of defeat of Hindu rulers and the beginning of Muslim rule in northern India. The Soorsutty formed the real and metaphorical line in this historic conflict.

The book also described Feroze Shah's canal projects, with frequent references to the Soorsutty as a historical river. Feroze Shah built an aqueduct over the Soorsutty, which is described as a small tributary of the Sutlej, to carry water to Firozabad, a city that he founded and named after himself.[82] In 1360, as Feroze Shah returned to Delhi from a military campaign, he heard that near Perwar (most probably Parwah in Himachal Pradesh, around 250 kilometers northwest of Saharanpur) was a hill out of which ran a stream that emptied into the Sutlej by the name of Soorsutty, beyond which was a smaller stream, called the Sulima. He cut through a large mound that intervened between these streams and drained the water of the Soorsutty into the small stream. This provided a perennial course of water to the city of Soonam, passing through Sirhind and Munsoorpoor.[83] This Soorsutty was in the foothills of the Himalayas. The British built yet another canal over Feroze Shah's canal there, known as the Sirhind Canal. The project was started in 1860 and completed in 1882.[84]

The nineteenth-century story of the Saraswati is enmeshed with history and myth. Historically, there seems to have been a river called Soorsutty, which Farishta referred to, that ran through the semiarid tracts west of the Yamuna and east of the Sutlej. There is also a mythical "lost" one, referred to in the Vedic texts as the Sarasvati.[85] Besides these were local folk tales of several lost rivers linked to the myths of the Saraswati. However, in the nineteenth century, while the British found a place called Sirsa (still there, a small town in Haryana by the River Ghaggar), there was no river called Saraswati or Soorsutty. British engineers found a minor tributary of the Sutlej called Sirsa or Soorsa, west of the Yamuna. They assumed this was a derivative of Farishta's Soorsutty, which too was described as a tributary of the Sutlej. Briggs, in his translation of Farishta's work, commented, "The Soorswutty is called the Soorsa in our maps."[86] Farishta referred to another small stream by the name of Soorsutty, in the foothills of the Himalayas, on which the British built the Sirhind Canal.

Most modern historical accounts of the Saraswati refer to the article "The Saraswatī and the Lost River of the Indian Desert," written by the surgeon and Indologist Charles Frederick Oldham, published in 1893. The article drew mainly from Friedrich Max Müller's *Sacred Books of the East*, published in 1879, for the Vedic references to the Saraswati.[87] C. F. Oldham traced the course of the "lost river" from the Himalayas to the Rann of Cutch. He, in fact, wrote two articles on the mythical river. The first, written anonymously, "Notes on the Lost River of the Indian Desert," was published in 1874, *before* Max Müller's volumes were published.[88] As a young surgeon in the 1860s, C. F. Oldham had surveyed the region from the Himalayan foothills of the Punjab to the Thar Desert of northwestern India. Drawing from these experiences, he was the first to situate the mythical river in this landscape of lost rivers and canals west of the Yamuna. Oldham's first article, which is not cited as much as his later one, was based partly on his geographical explorations and historical research of the region.

This article was also marked with a clear lack of fixity, in terms of both geography and etymology, representative of the several works of its genre by British engineers and explorers on the rivers and canals of the region. He referred to the "Sarsuti" quite freely as the Saraswati and quoted from Farishta's history to state that Feroze Shah had dug through the dead riverbed of the "Saraswati" and converted it into an irrigation channel.[89] It is not clear whether here he meant the Saraswati of the Vedas or was merely following the common belief that the Sarsuti was a derivative of the Saraswati. This ambiguity was partly due to the variety of sources he consulted. He referred to Farishta's history along with the thirteenth-century Persian text *Tabakat-i-Nasiri* and the fourteenth- and ninth-century Arabic texts of Ibn Battuta and Al-Balahduri, respectively. These Arabic and Persian sources helped him trace the medieval history of the Soorsutty. According to him, local traditions suggested that the Thar Desert of India was formed by the drying up of the Saraswati, which irrigated it. The dry bed of a large river, people believed, could still be traced from the Himalayas, through Punjab into Sindh then on to the Rann of Cutch. Merging his geographical knowledge of the landscape with his textual readings, Oldham rejected that suggestion and argued that the river could not have dried up, as there was no evidence of the mythical river bringing water to this drought-prone region. The region was arid even in the Vedic period, and it was much more likely that the rivers changed course, as they frequently did in the past in this region.[90] He noted that rivers called the Saraswati or Sursuti or Ghaggar carried very little water in medieval times or at present. There was no evidence that "these rivers ever contained much more water [in the past] than they do now."[91]

He opted for a much simpler explanation: that an existing major river was known as the Saraswati in Vedic times and the name had merely changed. In the first article published in 1874, which I am still drawing from here, he was also much less clear about the mythical Saraswati. He suggested that the name could have been a ubiquitous one that was used for the present-day Sutlej: "The missing river was not the Ghaggar, nor the sacred Saraswati, nor yet a mythic stream; but was no other than the well-known Satlej."[92] The Sutlej, of which the Soorsuttee, as referred to by Farishta, was a tributary, is the easternmost of the five rivers of the Punjab and the second major river west of the Yamuna. Researching through layers of historical texts, local folklore, and geographical studies, Oldham suggested that the Sutlej originally flowed through the course of the Hakra. The Hakra did not dry up because of a decrease in rainfall; its waters stopped flowing through its ancient bed and found their way through another channel. Therefore, according to him, the Saraswati was not a mythic river but merely an old course of the Sutlej.[93]

The Saraswati, therefore, had multiple identities in the nineteenth century, traced through archaeology, geology, and Indo-Islamic history. It was simultaneously a lost or a modern river whose name was merely lost, or several different rivers with the same name in a vast landscape, or a dried river that became a canal, or just a line through the desert. Despite this lack of fixity, C. F. Oldham established one possibility in his 1874 article: that the mythical lost river could be traced textually and geologically in the landscape of northwestern India. From this point on, local folklore and textual mythology could not remain independent of each other. Geology became, as we shall see, the foundation that bound these various channels of antiquity together.

While the battle lines of history appeared clearly drawn in the river that Muhammad Ghori crossed in the twelfth century, the river itself, as we have seen, appeared fragmented. The seismologist R. D. Oldham (son of the Geological Survey of India geologist Thomas Oldham), who also surveyed the river courses of the region, sought to provide geological clarity to these layers of texts, folklores, changing names, and shifting riverbeds. He studied the seismic changes in the Himalayan foothills and explained the difficulty in tracking the antiquity of rivers in a region where they have either moved frequently, were sometimes used as canals, or whose names and legends had moved. To make matters more complicated in this landscape of seismic changes, the River Indus itself was not always a fixed etymological category. Historians have shown that what came to be known as the Indus was a derivation of the Sanskrit word *sindu*, or *sindhu*, which could refer to *any* river.[94] During his seismological research, R. D. Oldham found several dry river channels, all of which led from within a few miles of the present channel of the Sutlej and ultimately joined the dry bed of the lost river when the Sutlej turned west to merge with the River Beas. He agreed with C. F. Oldham that the river known by the name of Sursootty could not have been the major river of the Vedas: "This lost river of the Indian Desert was none other than the Sutlej."[95]

In his paper on the rivers of the Punjab and Sindh, R. D. Oldham provided the first formal exposition of what I am referring to here as the natural-historical method (he called it the "historico-geographical study") that the British adopted there in the study Indian antiquity through rivers and canals. He suggested that myths had historical value, as "no tradition ever arose without some foundation in fact." However, as oral traditions often changed, they needed to be verified by historical accounts, such as those of Greek and Arab historians who had written on the geography of Sindh. Yet, as these too were often vague in geographical terms, they needed to be corroborated by the geological study of the landscape.[96]

Therefore, according to R. D. Oldham, the process of establishing authenticity involved moving from myths to history and then to geology. To demonstrate his point, he used the example of the eastern Narra, in Sindh. The eastern Narra also appeared to British engineers to be an old canal during excavation in the 1850s.[97] Popular folklore and historical narratives generally regarded the eastern Narra as the old course of the Indus, through which Alexander's troops sailed when they entered India around 326 BC. Alexander Cunningham referred to the same idea in his *Ancient Geography of India.*[98] However, Oldham believed that geology pointed to a different historical fact. He found that geologically it was difficult to establish that that river left its low-lying tracks of the Narra and moved to a higher ground, in its present course. This unfeasibility could become evident, according to him, only through the geological and seismological study of the landscape, which superseded erstwhile historical narratives: "Thus accepting prevalent ideas, without learning about the principles of physical geography, leads us to make such problematic conclusions."[99]

R. D. Oldham similarly surveyed the terrain through which the Indus flowed. According to him, there was historical and geographical evidence of the former existence of a river independent of the Indus that flowed to the Rann of Kutch, and this was the former course of the eastern Narra. He referred to Greek and Arab historians, all of whom gave vague descriptions of the geography of Sindh. He believed that Arab knowledge of the geography of Sindh was as poor as early nineteenth-century British knowledge of Central Africa or inner Tibet.[100] Therefore, he suggested, it was only with modern seismological knowledge that the true history of these lost rivers could be established. This method, of turning myths first into history and then situating them in the landscape, which developed through these traditions, became the model for future Indian engagements with geomyths.

In his search for geohistorical fixity in this land of shifting antiquities, Oldham turned to the *Chachnama*, a thirteenth-century Persian account of the Arab occupation of Sindh. British scholars had translated the text around the middle of the nineteenth century and considered it an "authentic" source for the history of the region.[101] The text appeared to support his geological insight that the eastern Narra was not the ancient bed of the Indus but a distinct river that ran alongside it.[102] According to him, it referred to the "waters of the Sakra" near the Indus, which he believed was the Narra in the past and was the same as the Hakra, the dry bed of a lost river in Rajputana.[103]

Oldham next tackled the question of the "the Saraswati of the Vedas" with his approach.[104] He seemed to treat this differently than he treated the lost river of the Indian desert, which he associated with the Sursootty and as an older reference to the Sutlej. A complex analysis of hydrographic changes, the gradient of the land, various old riverbeds, and mentions of names of rivers in Vedic and various other sources suggested that the Saraswati could have been an old bed of the Yamuna. He noticed that in Vedic hymns, the Saraswati was mentioned in the same passage as the Yamuna, and the slope of the land at the point where the latter entered the plains of India would have allowed it to branch off toward the west as well. He, therefore, placed the mythical river along the real one and suggested that the Yamuna, after leaving the hills, originally bifurcated into two courses. The one that flowed through Punjab was known as the Saraswati, and the one that joined the Ganga was called the Yamuna. This was possibly the hydrographic situation of the region when the Aryans came to India.[105] He concluded, "We may take it as proved that there have been great changes in the hydrography of the Punjab and Sind within the recent period of geology . . . by which a large tract of once fertile country has been converted into desert."[106] The deep past of the region was established through this search for geological and historical antiquity. The myth of the Saraswati became credible because it blended history and myth with geohistory.

In his later article, which he wrote from England after his retirement in 1890, C. F. Oldham depended heavily on Max Müller's *Sacred Books of the East*, which were translations of Vedic texts, and combined them with the seismological evidence and land and canal surveys.[107] Max Müller had suggested that the Vedas mentioned a major river named Sarasvati (yet another version of the name), which later Sanskrit texts described as having disappeared in the desert. In Max Müller's translations, the Sarasvati was annotated with nineteenth-century geological and geographical ideas that sought to provide a natural shape to Vedic geographical references.[108] He suggested that in the Vedic times, the Sarasvati was a major river, the easternmost of the rivers of the Punjab, and formed a *natural* border of India, "an iron gate, or the real frontier against the rest of India."[109]

C. F. Oldham brought these Vedic references to the geographical backdrop of the western Yamuna region by integrating the entire genre of nineteenth-century antiquarianism that traversed the natural and historical, and combining it with information on water levels and aridity, as well as legends. He created a composite and authentic narrative of the mythological river. Oldham sought to trace a singular and continuous river as suggested by Max Müller's reading of the Rig Veda. There the Sarasvati emerged from the Himalayas and ended in the sea, but it lost its way in the desert in subsequent years. In this scheme, the different Saraswatis, in names and forms that had been unearthed over the nineteenth century, came together as he identified the Soorsootty of the Himalayan foothills, through which Feroze Shah dug his channel, as the early course of the Vedic Saraswati. In the plains, he identified the Ghaggar or Soorsuttee, which Colvin had traced, as the continuity of the same river. The various Saraswatis, Soorsas, and Soorsoottys became one in this grand collaged cartography of myths, nature, and natural history.

Oldham finally asserted that the Ghaggar "was formerly the Saraswati" and that "the famous fortress of Sarsuti or Saraswati was built upon its banks."[110] As the Ghaggar itself was a dead river, he traced the Saraswati through several dead riverbeds and ancient canals found in the region during the nineteenth century, including the Chitrang, which was one of the canals that Feroze dug, possibly over a dead riverbed.[111] He was uncertain why the Saraswati had lost its original name and acquired a different one (Ghaggar). He speculated that it "may have been owing to some change in its course in comparatively modern times."[112] This river then moved in a southwesterly direction, first to become the Hakra, and then ran through the old riverbed of Narra (which too, as I observed before, was a medieval canal) in Sindh, which then moved toward the sea.[113] A contented Oldham concluded, "The course of the 'lost river' has now been traced from the Himalaya to the Rann of Kach."[114] He produced the first natural-historical map of the mythological river in his later article. In dotted lines, he mapped the imagined bed of the Saraswati from the Himalayan foothills to the sea (fig. 1.4). The mythical Saraswati became real.

This mapping of an imagined river and landscape became the foundation for later suggestions about the true course of the Saraswati. It also provided the blueprint for future projects that similarly embedded textual sources with geological evidence.[115] Satellite imaging, which appears to have traced the lost river in the desert, has backed these claims.[116] In 2016, the provincial government of Haryana even pumped water through the dead river channel in the desert.[117] This new river was dug over the landscape of the canals of Zabita Khan and Feroze Shah. The physical entity of the Saraswati River is now etched deeply in the arid landscape west of the Yamuna. The mapping of the Saraswati

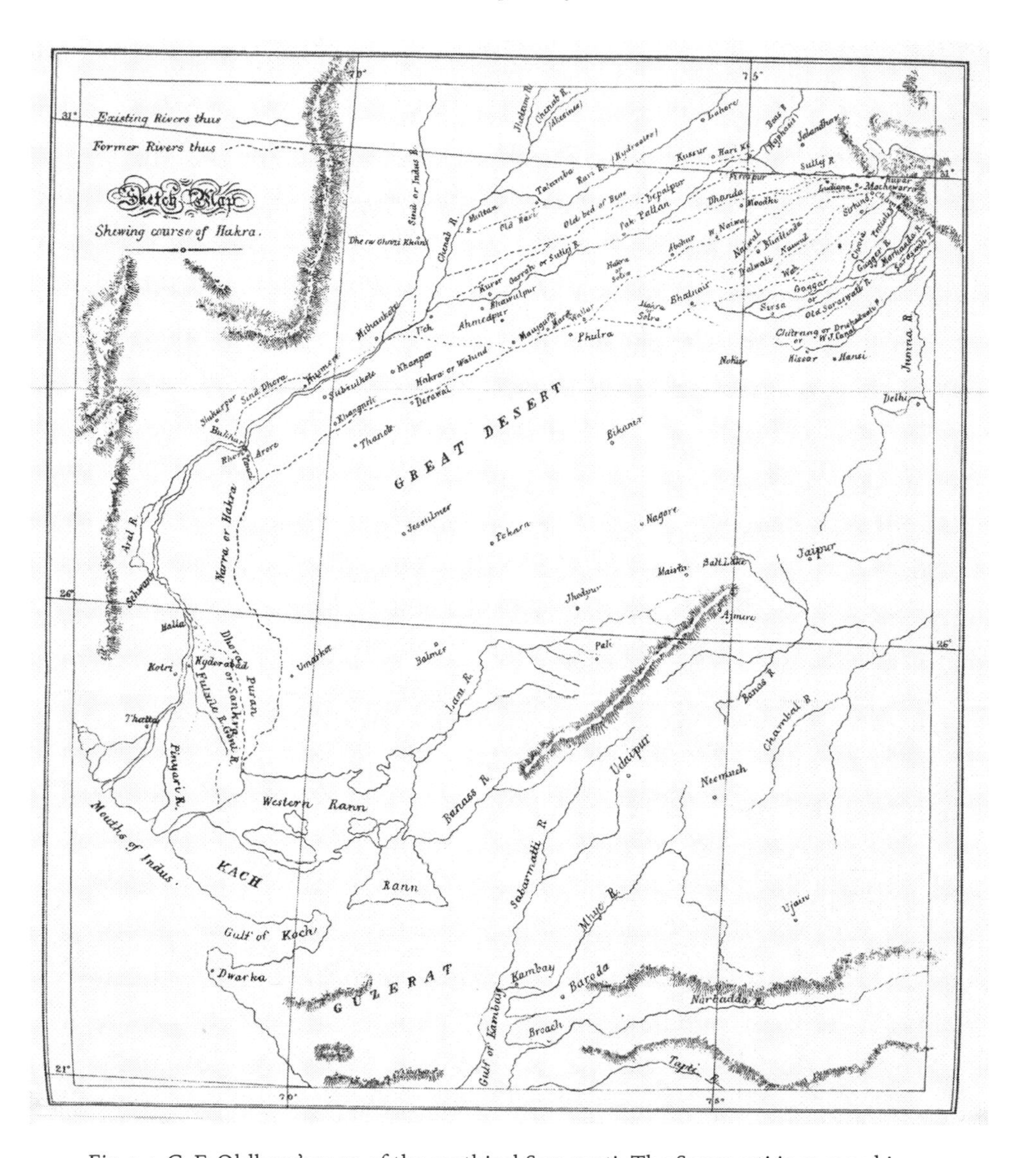

Fig. 1.4. C. F. Oldham's map of the mythical Saraswati. The Saraswati is mapped in dotted lines. C. F. Oldham, "The Saraswatī and the Lost River of the Indian Desert," *Journal of the Royal Asiatic Society of Great Britain and Ireland* (January 1893), after p. 76. Image copyright of the University of Manchester.

marks an important chapter in the imagination of India as a deep Hindu landscape.

As the geological past of the subcontinent deepened, so did the antiquity of the Saraswati. With the emergence of theories of plate tectonics in the 1930s, geologists such as Guy E. Pilgrim and Darashaw Nosherwan Wadia (both members of the GSI), suggested that in the prehistoric period, all the three major river systems of northern India, the Indus, Yamuna-Ganga, and Brahmaputra, were once part of a single river system, which rose in Assam in the northeast and flowed along the foot of the Siwaliks, westward, draining finally into the Arabian Sea. Pilgrim named the main river the "Siwalik river."[118] The Ganga and the Yamuna flowed in the opposite direction of their current courses, from east to west, as tributaries of this grand prehistoric Siwalik River. They argued that these rivers changed direction because of plate movements. Wadia found "both physical and historic grounds" to believe that the Yamuna discharged its water into the Indus system, through the now-dead riverbed of the Saraswati.[119] This tended to confirm R. D. Oldham's seismological speculation that the Saraswati was an old riverbed of the Yamuna. In the wide-open and barren landscape west of Delhi between the Indus and the Yamuna, Wadia and others saw the geographical and oral traces of this prehistoric river system, which people remembered in the form of myths of dead riverbeds.[120] Subsequently, the Saraswati appeared as *that* "prehistoric river" of India, holding all the clues to the birth of Indian civilization.[121] On the banks of the Yamuna, as history met prehistory, one mythical river merged into another.

Ghosts of the Yamuna: Prehistory Comes to the Doab

In the excavations for the canal of Zabita Khan, the search for antiquity moved between texts, monuments, and the landscape. In the landscape, the possibilities of the past, as they were aligned with the history of the earth, appeared unlimited. While digging the canal, British engineers became aware of the dynamic nature of river courses in the region, which indicated shifting layers of antiquity. In 1833, Cautley was clearing the old cut of the canal in Behat, 20 miles north of Saharanpur in the foothills of the Himalayas. He found the hilly landscape crisscrossed with torrential ravines, waterfalls, and beds of the old canal. Cautley observed that the mountain torrents changed their courses frequently: "Such things are in constant progress, and things of annual occurrence!" The local zamindar (landlord) told him that the track on which the new canal was being built used to be adjacent to the river in his childhood and was used for paddy cultivation. The river had now moved farther away.[122]

While digging the earth in Behat, Cautley discovered an ancient town buried seventeen feet below the surface. He believed that the slow and gradual deposition of silt from shifting earth of the lower hills had buried the city.[123] He initially thought that the ruins were the remains of an old canal works.[124]

As he excavated farther around the site, he realized that this was an ancient settlement submerged by the torrents. The ancient canal had probably enlarged one of the ravines, called Behat khala, which drowned the town. As he studied the landscape and sought to dig the canal for water two miles north of Behat at a depth of 30 feet, Cautley reached a bed of shingles, common to all the riverbeds of the region. This was an old riverbed buried under 30 feet of soil. It became apparent to him that the town was at the level of the former riverbed, and the enormous discharge of soil from the hills through the torrents had raised the entire region around it. As the digging continued, he found old riverbeds in other places as well, all around 30 feet under the surface. He believed that either Feroze Shah or others had built the old canal on top of the ancient riverbeds.[125] A submerged civilization began to emerge.

Behat initially formed part of a classical archaeological study. Cautley described the buried town as the "Oriental Herculaneum," after the ancient Roman city buried by the eruptions of Vesuvius.[126] In Behat, he found fragments of old pottery and coins that were, in fact, the first major finds of Indo-Scythian coins.[127] The numismatic discovery was significant because this occurred at the time of the British discovery of Indo-Greek heritage. Prinsep, the Orientalist scholar, as the secretary of the Asiatic Society of Bengal (1832–39), oversaw one of the most productive periods of numismatic and epigraphic study in nineteenth-century India. Between 1833 and 1838, Prinsep published a series of papers based on Indo-Greek coins and his deciphering of Brahmi and Kharoshthi scripts.[128] The Behat coins prompted Prinsep's first paper on Indo-Greek numismatics, published in 1834, where he connected the Behat find with the Greek coins (fig. 1.5).[129] He consulted Wilson, the Orientalist, who could not find any textual reference to an ancient town in the region. He therefore presumed that the site in Behat could have been a Buddhist monastery that was abandoned and gradually destroyed when the sect was persecuted around the third or fourth century AD and the city was subsequently submerged by the torrents coming down from the mountains.[130]

However, it soon became evident that various strata of the past had collapsed in Behat. These classical archaeological finds and the linking of the Indian past with Greco-Roman history were soon set next to the discovery of prehistoric remains and their associated questions of the origins of humans in the Gangetic Plain. Among the ruins of Behat, Cautley also found fossilized teeth of alligators (called *magar* by the locals) and crocodiles (*gharial*). There were also fossils of smaller mammals, including a horselike creature, rats, and fish. There were other fossils that he could not identify.[131] Cautley recollected

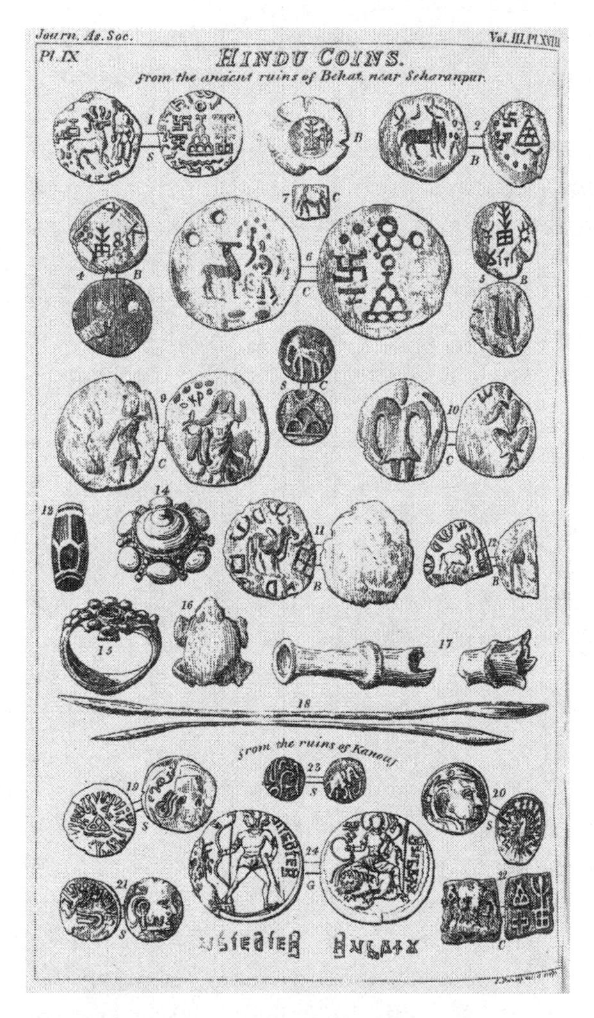

Fig. 1.5. Behat coins. James Prinsep, "Note on the Coins, found by captain Cautley, at Behat," *Journal of the Asiatic Society of Bengal* 3 (1834): 226–27. Image copyright of the University of Manchester.

that Feroze Shah, while digging his canals around the same region had also discovered fossils and indeed was the "original discoverer of the fossil remains."[132] Cautley quoted from Briggs's *History of the Rise of the Mahomedan Power* that in 1360, Feroze had employed 50,000 laborers to cut through a hill to connect the canal to the Sutlej River near Perwar. As they dug through the hill, they came across "bones of Elephants and men."[133]

Although it is not clear whether any of the fossils found by Cautley at Behat belonged to humans, the search for the antiquity of the canal now assumed prehistoric significance. As the editor of the *Journal of Asiatic Society of Bengal*, Prinsep urged Cautley to now excavate Feroze Shah's canals for geological knowledge, as it could lead to more such fossils, taking a lead from the latter's own efforts: "It is seldom that a geologist can command the aid of fifty thousand men to open a section of the Himalayan strata to this view."[134] Feroze Shah's canals were now part of the new journey into the deep antiquity of the Ganga-Yamuna Doab.

The caves of Southern France gave new meanings to deep time around the same time that these fossils were discovered. Until the 1830s, human history was believed to be only as old as human records and monuments, such as the Egyptian hieroglyphs and ancient ruins. The discovery of fossilized human bones began to inscribe human antiquity into the history of the earth. The study of human origin through fossils not only prolonged the era of the human past beyond biblical accounts; it also linked human history with geohistory. From the 1830s, deriving from his discovery of human fossils among those of animals and fragments of pottery, Paul Tournal suggested for the first time that human antiquity appeared to be of similar age as those animals and therefore preceded biblical depictions.[135] He almost single-handedly invented prehistory, although not the term itself.[136] His claims, however, did not go uncontested. Major debates unfolded among naturalists and theologians in Europe over the next few decades about their reliability.[137]

These debates took place in Europe as more fossils were being found around the Yamuna. Farther south of the Himalayan foothills, in the Gangetic Plain, the rivers posed different challenges to the engineers. The riverbeds needed to be cleared to build the canal and receive water from the Yamuna. Silt deposits, which the Yamuna had carried over long periods, in the shape of raised *kankar* beds called "flag" on the banks and at the center of the river, proved the problem. Initially, because the deposits were uneven, the water flooded various parts of the canal.[138] The eventual clearing of these beds enabled "cotton boats" of larger tonnage to reach all the way up to Agra, carrying bales of cotton from northern India to Calcutta to be shipped to England.[139] In 1832, Captain Edward Smith, a member of the Bengal Engineers, was in charge of clearing the *kankar* beds of the Yamuna for the canals between Allahabad and Agra. Traditionally, when the water was low, men belonging to the *mallah* (traditional boatmen caste) community would break up and clear these beds for navigation. Local people asserted that these *kankar* beds were never found on the sites of former excavations, which suggested to Smith that these were all of prehistoric formation.[140]

As the *mallahs* started dredging the beds by hand and with simple spear-like tools, Smith found in these deposits mammal fossil bones in "great abundance."[141] He identified some of them as belonging to elephants, camels, alligators, buffalos, and horses. Others, he believed, were remains of human bones. He sent them to the Asiatic Society in Calcutta.[142] As the dredging continued, Smith found more fossils, some so large that they could not be carried to Calcutta.[143]

In 1834, Edmund Dean, an army sergeant employed at the Yamuna Canal project, was similarly clearing the alluvial deposits in the ancient canals and the river.[144] Once the shoals and rocks were removed, he found fossils of what he thought were "human bones."[145] They were in such a perfect state that he even asked a local physician to examine them. There were 16 pieces in total; some he believed belonged to a woman or a child, others to an adult male. Apart from these, he found an elephant's tooth, jaws, and teeth of other animals; the latter he speculated belonged to the "scarce monster," an elusive species of crocodile known as the "Bhote [ghost] of the Jamna."[146] Crocodiles in the nineteenth centry roamed the Yamuna freely around Delhi.[147] Dean believed that the Yamuna water had some "peculiar quality" to fossilize these bones.[148]

The fossils appeared at different sites along the river. John Leslie, a surgeon with the EIC, identified fossil remains on the banks of the Yamuna in the 1820s. Leslie sent a femur bone, discovered in the bed of the river near Calpee (Kalpi, an old town along the Yamuna) in 1828, and a portion of elephant tusk to the Asiatic Society of Bengal.[149] Robert Barclay Duncan, the resident surgeon at Calpee, also found fossils bones believed to be from elephants, camels, horses, and, in two cases, even humans.[150] These various fossils came to be known collectively as the "Jamna fossils," and they were the first major discovery of fossils in India (fig. 1.6).

The fossils found by Smith and Dean were sent to the Asiatic Society in Calcutta, where Prinsep examined them.[151] In a note titled, "Occurrence of the Bones of Man in the Fossil State," Prinsep connected the Jamna fossils to recent deliberations in Europe around Tournal's discoveries of human remains in the caves of Southern France.[152] The idea, Prinsep noted, "that man was the companion of animals now considered extinct and fossil" was gaining ground. He therefore encouraged others associated with the EIC and the Asiatic Society in India to explore the human and animal fossils of the Doab region. He believed that the alluvium of the Doab might prove to be geologically contemporaneous to the rocks and mud of the caves of Southern France.[153]

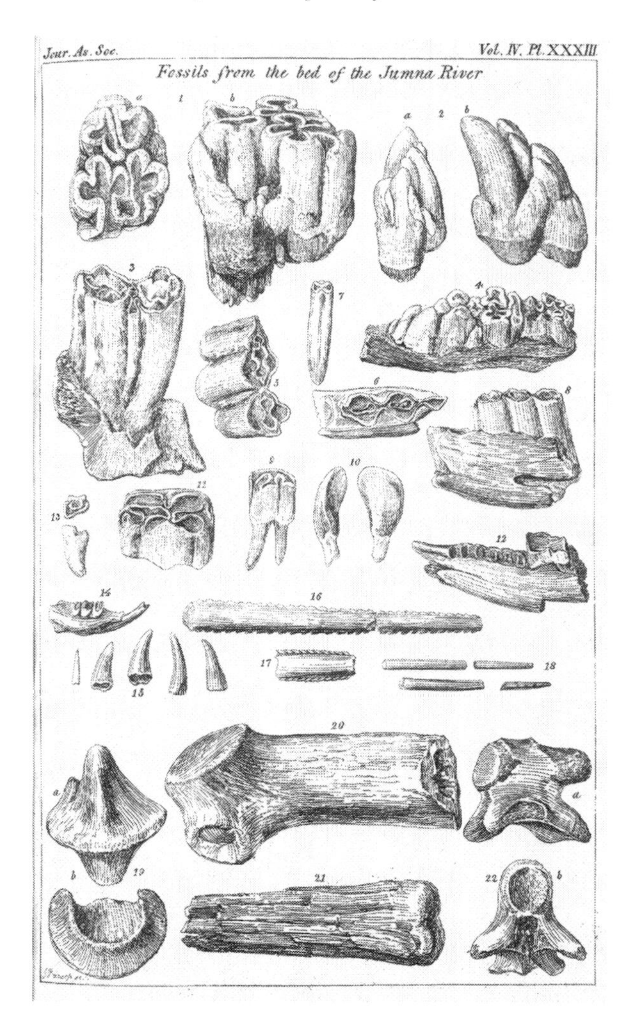

Fig. 1.6. Fossils of the Yamuna. *Journal of Asiatic Society of Bengal* 4 (1835), follows p. 506. Image copyright of the University of Manchester.

The deliberations around the Yamuna fossils, discovered at the time of Tournal's radical propositions, were centered on the question of whether they belonged to humans. Although it remained unclear whether the fossils found in the Yamuna indeed were human, the potential for prehistory to define the antiquity of India began to be appreciated. In his note on the Jamna fossils, Prinsep commented that the fossils of the Yamuna and the Doab now promised a new history of the region:

> We have dwelt at some length on this novel subject in hopes of drawing the attention of our Indian geologists more zealously to prosecute their investigation of the new field of organic remains now opened to their labours in the clay of *Doab* and the banks of the *Jamna*. Should it be proved that the bones of man are there really imbedded, . . . it will form a strong and very important link of connexion between the state of things at two distant epochs of our globe, now distinguished as the recent and fossil periods.[154]

We leave this chapter at the onset of this new history in the Ganga-Yamuna Doab. As we shall see in the next few chapters, the fossils found in the Yamuna and the Himalayan foothills ignited questions of deep human past in the Indian subcontinent, which transformed debates around Indian antiquity. For the moment, we can appreciate that the canal of Zabita Khan allows us to see how on the banks of the Yamuna, deep history redefined existing historical imaginations. The digging for the canal in Abbéville, which excavated the flint tools for Boucher de Perthes; or the Somerset coal canal, where William Smith found stratified distribution of fossils; or even the caves of Southern France, where Tournal found human remains, did not have the same multiple historical significances beyond the discovery of fossils and prehistory. The Doab, however, was the site of simultaneous explorations of Indo-Islamic history, classical archaeology, and prehistoric geology that reflects this particular reconfiguration of history.

Conclusion

This history of the canal of Zabita Khan highlights the modes through which the historical and the mythological turned into the geological. It provides a unique opportunity to understand how antiquity appeared to be inscribed in the natural landscape. The Indo-Gangetic Plain has often been seen as the bedrock of Indian civilization and antiquity. There are several reasons for this. To begin with, this was one of the primary sites of the British discovery of Indian antiquity. This antiquity was located simultaneously in Sanskrit and Indo-Islamic texts and in its monuments. More importantly, this antiquity was then traced in the landscape itself. The entrenchment of history with nature made these myths and their associated antiquities appear timeless, and the region as the foundation of Indian antiquity. The mythical rivers appeared to be real. The desolate landscape suddenly appeared as witness to a prosperous past. The canals and rivers that existed in the collective memories of peoples appeared to be evidence of prehistoric rivers.

The discovery of the predominantly urban Indus Valley civilization in the Punjab and Sindh in the early twentieth century added yet another dimension to the prehistoric past of this region. Discussions on the emergence and decline of this civilization have been defined by deep naturalistic interpretations as well,[155] through geohistorical analyses of droughts, floods, and changing climatic patterns.[156] On this historical question, scientists and geographers have participated equally with historians, as they have done for the mythical Saraswati.[157]

We have seen how the deep past came to the surface of nineteenth-century India. While in Europe, the geological, archaeological, and historical imaginations also concurred, the sites rarely overlapped as they do at the canal of Zabita Khan. Each of the episodes discussed in this chapter—the Doab Canal project, the British discovery and restoration of Indo-Islamic architecture, the genesis of colonial archaeology, the discovery of the Saraswati, and the search for prehistory—enjoys distinct historical trajectories within existing academic and public discourses. This fragmentation is due to the emergence of distinct disciplines, which allows antiquities to be traced only through specific lineages. When narrated together, they enable us to understand the infusion of geological antiquarianism within the historical one.

Ancient Alluviums

The Geology of Antiquity

Zabita Khan had built a pleasure garden known as Farhat-Baksh in Saharanpur, close to the Yamuna in the foothills of the Himalayas.[1] The East India Company acquired it in 1818 during the construction of the Doab Canal.[2] As construction work started, Saharanpur, situated near the old cut of Feroze Shah's canal, became headquarters of the eastern Yamuna canal works.[3] Khan's derelict pleasure ground became the botanical garden and the museum of natural history of the EIC and an important site for Indian fossil research in the nineteenth century. Saharanpur, located in the foothills of the Himalayas, provided a vantage point from which to make the thesis of the geological evolution of the Indian subcontinent. From there, Falconer and Cautley linked the Himalayan and Yamuna fossils with Himalayan natural history. This eventually contributed to the theory of the formation of the Gangetic basin and the Himalayas and of the origin of humans in the Indian subcontinent. Our story moves in this chapter, along with our chief protagonist, Hugh Falconer, between three different sites, Saharanpur, London, and Calcutta.

Along this journey we will delve into how the colonial and Orientalist antiquarianism of India became embedded within a geological narrative of the Indian subcontinent, a narrative that, in turn, presented a distinct imagination of human antiquity. It also dislocated discussions of human origins from Europe to the imagined landmass of Indo-Africa. The Himalayan and Gangetic landscapes were sites of the first major discovery of fossils in India in the nineteenth century. These fossils defined theories of the emergence of primeval humans in India as well as of the geological formation of the entire subcontinent and its presumed links with Africa. In the late nineteenth century, several geologists suggested that the river basins of India and Africa were sites of human evolution. The theories drew not just from studies of fossils and paleontology but also from a range of Orientalist and Indological traditions that were simultaneously deliberating upon the origin of human civilization in Asia. This allowed scholars to imagine unique connections between the historical and natural traces of evolution. These ideas coexisted with Darwinian theories of natural selection, and in their distinct trajectories, they provide us with a new understanding of the historicity of nature.

It is possible to suggest that in this story, the mountains came later. This is a twofold inversion of the Braudelian proposition "mountains come first," which sees mountains as the foundational geohistorical category of the Mediterranean world.[4] On the one hand, the scientific consensus among geologists in the late nineteenth century was that the Himalayas were a relatively recent formation and human life in the ancient alluviums of prehistoric river systems of the Indian subcontinent preceded it and retained signs of that geological and ecological transformation. Geologists suggested, as we shall see, that the Himalayan mountains rose from the bed of the great ocean of Tethys, remains of which could still be found in the lakes of central Tibet. The Indian geologist D. N. Wadia explained in 1919, in his textbook on the geology of India, that the drainage systems of the Gangetic rivers were "*antecedent*" to the formation of the Himalayas, although the latter was now their source of water.[5] Therefore, the prehistoric ocean and the land around it were their primary subjects of investigation into the prehistory of India. The other suggestive inversion is that the British approached the Himalayas from their historical and geological experiences of the Doab plains. The Gangetic Plain, the primary site of British natural-historical encounters in India, as discussed in the previous chapter, shaped their imagination of the Himalayas.

The botanical garden at Saharanpur, located between the Yamuna and the Ganga, at the feet of the Himalayas, appeared to be the perfect site to study the natural history of the Indian subcontinent. As Falconer's friend and biographer, Charles Murchison, pointed out, "It [Saharanpur] is thus most favourably situated as a central station for Natural History investigations; the rivers, plains, forests, and hills teeming with life in every shape, and the range of elevation combining, within a short distance, the features and productions of tropical temperate and alpine regions insensibly blended." There Europeans could find seclusion from the plains of India to conduct their research. They could yet call upon, when necessary, the "intelligence, docility, and exquisite manual dexterity of the natives, backed by their faith in the guiding head of the European" in their pursuit of natural history.[6]

Besides the necessary seclusion of the mountains and the essential obedience of the locals, geographically too Saharanpur and the garden were crucial to the study of Himalayan natural history. George Govan was the first superintendent of the botanical garden, followed by John Forbes Royle in 1823. Royle was a British botanist born in Kanpur in north India and entered the service of the EIC as an assistant surgeon.[7] He remained the superintendent of the garden for nearly ten years, during which time he studied botany and geology and amassed large collections of plants, fossils, and cultural artifacts from the mountains using indigenous trading and pilgrimage networks across the Himalayas. As he once wrote to the Calcutta-based botanist Nathaniel Wallich, "By means of the Northern Merchants on their return from the Hurdwar fair I shall be able to obtain roots and plants of the saffron which . . . had some years since been introduced from Cashmere into . . . [illegible] near Subathoo and from thence into Dehra in the Dhoon."[8] Royle regularly sent his own Indian "collecting parties" to the snow-capped Himalayas to bring him rare and exotic plants.[9] In Royle's words, this enabled him to "pick up one or two of the lost links in the history of the science."[10]

Royle established a museum of natural history at the Saharanpur garden, where he collected fossils from the coalmines of Burdwan, fossilized antelope skulls from the Himalayas, fossil bones from Tibet, Cautley's fossils from Behat, and Alexander Gerard's (elder brother of James Gerard) shaligrams (ammonites) from the Spiti Valley, a desert region in the central Himalayas. When he returned to England in 1831 to take up a professorship at King's College, Royle completed his book on the natural history of the Himalayas.[11] Here he referred to the history of Himalayan fossil discoveries. He wrote that the Gandak River basin in Nepal had "long been known to bring down Fossil Ammonites, which are called Saligrammi and are much esteemed by the Hindoos." He also wrote about fossils known to locals as *bijli ke har*, or "lightning bones," which were used as medicine. His *Illustrations of the Botany and Other Branches of the Natural History of the Himalayan Mountains* included images of Himalayan fossils collected by various British explorers such as W. S. Webb, George William Traill, Falconer, and Cautley.[12]

Falconer arrived in Saharanpur in 1830, attracted by Royle's collection of Himalayan natural historical specimens. He became Royle's deputy at the garden in 1831 and in the following year, acquired full charge of the garden after Royle left for England. There he met Cautley, who was based at the headquarters of the canal works. Cautley showed him the "black cylindrical fossils" that he had come across during his canal excavations.[13] Falconer and Cautley were intrigued by Farishta's account of Feroze Shah's discoveries of "bones of giants . . . in the hills in which the Sutlej took its origin."[14] They began their explorations for Siwalik fossils from Saharanpur. They also appointed "Hindoo collectors" to identify and gather fossils from the hills.[15] Falconer himself set out in search of fossils in the Himalayan foothills and returned, as he reported to Royle, "loaded . . . with noble fossils of the monsters of the deep!"[16] In such a pursuit, the company's garden in Saharanpur became the repository of the new knowledge about the Himalayas and the Gangetic Plain and thereby a distinct institution of colonial natural history, rather than a continuation of Mughal gardens, botanical traditions, or aesthetics, as suggested by Richard Grove.[17]

I will begin with an explanation of how British geologists proposed the theory of the geological evolution of the Himalayas and the Indo-Gangetic Plain based primarily on fossil discoveries in the Siwalik Hills, the Tibetan plains, and the river basins of the Indian subcontinent. I will then show how that natural historical narrative was combined with existing Orientalist and Indological ideas of Indian antiquity, leading to the emergence of the grand natural-historical account of the Indian subcontinent. This marked a transformation of the history of fossils in India, where the geological evolution of the subcontinent became a natural-historical account of Indian antiquity.

Ancient Alluviums: The Natural History of the Himalayas

British travelers and prospectors started a distinct tradition of Himalayan exploration in the early nineteenth century. They rewrote the geography of Himalayas derived previously from Sanskrit and Tibetan sources and those of Jesuit missionaries based in Tibet.[18] These new explorations were mostly in search of trade routes through the mountains into Central Asia. For example, Samuel Turner's account of his journey to Tibet (1800) was driven by the ambitions of the governor of Bengal, Warren Hastings, to promote British trade across the Himalayas.[19] Webb's survey of the Kumaon region in the western Himalayas was motivated by mercantilist fantasies of finding precious metals in the depths of the mountains: "It cannot be doubted, that the mountain districts contain the precious metals, from the well-known fact, that the lands of almost every mountain stream are assiduously washed for gold at the points, where their rapidity diminishes."[20] These journeys marked the first modern mapping of the Himalayan routes and the emergence of a new geohistory of these mountains. William Moorcroft, a veterinary surgeon with the EIC, visited Lake Manasarovar (a lake in the remote Inner Himalayas), which was revered as a sacred site by Hindus.[21] Moorcroft's notes became a vital source of information for future explorations of the central Himalayas undertaken by Alexander Gerard, Webb, and Falconer.[22]

These early Himalayan explorations also led to the British discovery of Himalayan fossils. Webb collected fossils sold in local markets as religious and medical artifacts across the central Himalayas and sent them to the Asiatic Society of Bengal in Calcutta.[23] In 1821, during a trip to the central Himalayas bordering China at a height of 16,000 feet, Alexander Gerard found the flat terrain "studded with ammonites."[24] Falconer himself went on an expedition to Kashmir and Tibet in 1837–38. Starting from Kashmir, he crossed the high mountains around the source of the River Indus and then passed through snowfields before reaching the bleak and barren central Tibetan plains. He traveled through remote parts of Tibet, near mount Muztagh. On his way back, north of Nahun (a small kingdom in Himachal Pradesh), he found an embedded mastodon's head and collected "cartloads" of fossilized ivory. The locals seemed to be familiar with some of these bones and referred to them as *"rakass* [*raksas*] *ki har"* (bones of demons).[25] Falconer found fossils throughout the Himalayan foothills that appeared to belong to the same age. This led him to suggest that the entire range of foothills from the origins of the Indus to that of the Ganga had a similar formation.[26]

As the foothills of the Himalayas appeared to be a distinct geological region, they also acquired a new name. Cautley noted a lack of local names for the almost uninterrupted range of mountains between the Rivers Sutlej in the west and Brahmaputra in the east, a stretch of almost 1,200 miles that provided the bulk of the Himalayan fossils. He was aware that the local name for a smaller segment of these hills in the western parts was Shibwalla, which signified it as the residence of Lord Shiva. Deriving from it, he proposed the name "Sevalik," for that entire range of the foothills.[27] Sevalik, or Siwalik, as this range came to be known, was a geological term used to refer to its fossils and geological formation.

The Alps have been paradigmatic to the history of European geology.[28] From the end of the eighteenth century, a long line of "Alpine geologists," starting from Horace Bénédict de Saussure to Eduard Suess explored the complexity and "riddle" of the formation of the Alpine mountains. Almost all leading geologists in the nineteenth century—Lyell, Adam Sedgwick, Roderick Murchison, and Henry De la Beche—explored the Alps. They believed that the study of these mountains would address the main questions about the formation of the earth, contributing to various explanations from catastrophic elevations, plutonic upwelling, to continental uplift.[29] This Alpine orientation is also evident in the literature on the history of mountain geology, which overlooks the role Himalayas played in defining geological theorizations.[30] The Alps likewise loomed large over the geological explorations of the Himalayas in the early nineteenth century.

Geologists initially compared the Himalayan fossils to the fossils found in the tertiary formations of Europe. As Himalayan explorers found increasing varieties of fossils that did not conform to the Alpine ones, a new scheme of geological theory began to emerge around the formation of these mountains and, in turn, of the Indian subcontinent itself. Based on the discovery of Himalayan fossils, J. D. Herbert, a captain in the EIC army and assistant to the surveyor general of India, wrote the first detailed account of the organic remains of the Himalayas. He questioned the wisdom of comparing these fossils, which were found "in widely separated localities, in climates sometimes directly opposite" the European ones.[31] Herbert had previously surveyed the Himalayas to determine the heights and positions of the peaks.[32] He suggested that although the rock formations of the Himalayas were similar to the mountain ranges in Europe, the organic remains within these rocks were different. He criticized the position "hastily" adopted by European geologists that all mountains had uniform formations based on their Alpine and British experiences.[33] Although Herbert did not provide any definite theorization about the formation of the Himalayas, deliberations on which remained inconclusive until the theories of plate tectonics were accepted more than 100 years later, he suggested for the first time that the Himalayas had a distinct geological formation.

Falconer's theories of the formation of the Himalayas were based on these early propositions. He and Cautley undertook extensive fossil explorations in the Siwalik region from 1834. They found fossils of several new species, such as the extinct four-horned giraffid. They named it *Sivatherium*, after Lord Siva, who according to local myths resided in these mountains.[34] Falconer and Cautley also discovered fossils of the mastodon, stegodon, rhinoceros, giraffe, camel, and hippopotamus.[35] Cautley discovered remains of a "camelidae," which proved, he argued, that camels lived in India at the same time as the *Sivatherium*, rhinoceros, and the hippopotamus.[36] Falconer and Cautley depended on local intermediaries for collecting these fossils. The raja of Nahun possessed a large mastodon molar known as the "Tooth of Deo" and advised Falconer where to search for similar specimens. On one occasion, his advice helped Falconer to discover 300 fossilized bones in a day.[37] W. E. Baker, an engineer who assisted Cautley in the canal project, also found fossilized heads of elephants in the region.[38]

Along with the Himalayan and Yamuna fossils, explorers and geologists found fossils in all the major river basins of the subcontinent. They found them in the estuary of the Narmada River in western India and the Irrawaddy basin of Burma in the east. In Burma, the British civil commissioner in Rangoon and envoy to the court of Ava, John Crawfurd, traveled along the banks of the Irrawaddy River with the naturalist Nathaniel Wallich in the 1820s. They were surveying the region for forest resources when they found fossils. As the locals found out about their interest in fossils, they brought them several bones and shells that they had collected. Keen to identify the location of their find, Crawfurd and Wallich climbed a hill about 150 feet above the level of the Irrawaddy. They found some shells near the top. After some digging, they came across a bed of moist, blue clay that contained a large quantity of fossilized shells; some were broken, but most were intact.[39] Crawfurd and Wallich soon collected fossil bones of several ruminant animals, tortoises, and alligators from the Irrawaddy valley. The most numerous and remarkable bones seemed to belong to a mammoth. Crawfurd later donated them to the museum of the Asiatic Society in Calcutta.

Several of these early discoveries, like those in the Yamuna and along the Irrawaddy, were unplanned. In the 1830s, a lovelorn Austrian member of nobility, Carl von Hügel, was wandering around the ancient ruins of western India. He came across surgeon Charles Lush of the Bombay medical establishment, who had found fossil shells on the small island of Perim (now known as Piram) in the Gulf of Cambay, just off the coast of Kathiawar, close to the mouth of the Narmada River. Lush described to him how he found the fossil shells embedded in the building stones in the local villages.[40] He showed him fossils of a mastodon or elephant, a head of a boar, and a "rongeur" (rodent).[41] Hügel believed that the best evaluation of the fossils could be made either at Calcutta or in Saharanpur, where there were similar fossils from other parts of India for comparison. They decided to send the bones and shells to Calcutta to the Asiatic Society of Bengal.[42]

Hügel decided to visit Perim island himself. George Fulljames of the Bombay Engineers also became interested in the Perim fossils after hearing reports of "bones, turned into stones" and joined him. Fulljames extended the investigation into the entire Kathiawar Peninsula. He found mammal fossil bones at Chotah Gopnat Point at the mouth of the Narmada River at the Gulf of Cambay (fig. 2.1).[43] Although he could not investigate them thoroughly at the time, he believed that the fossils at the mouth of the Narmada were similar to those found in the riverbeds of northern India. As they collected more fossils on Perim island, Hügel realized that they were strikingly similar to those recently discovered in the Narmada River valley in central India. He believed "a vast field has thus been thrown open, for discovery and research."[44]

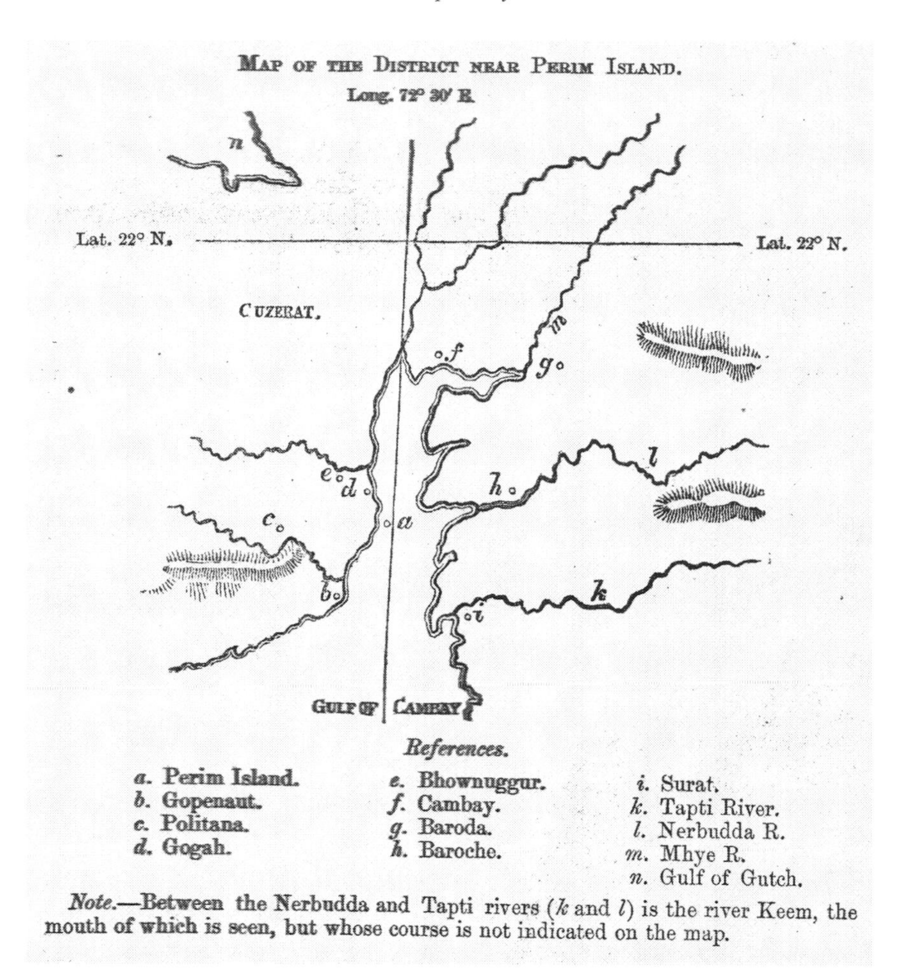

Fig. 2.1. Map of Perim Island at the Gulf of Cambay. Perim Island is marked as "a." Charles Murchison, ed., *Palaeontological Memoirs and Notes of the Late Hugh Falconer* (London: Robert Hardwicke, 1868), 1:392.

That "vast field" was established through a comparative study of the various fossils found in the different river valleys of the Indian subcontinent. A consensus gradually emerged among British naturalists in India from the late 1830s that fossils found in the Gangetic, Narmada, and Irrawaddy River valleys were interlinked in remote antiquity. In other words, these fossils belonged to the same group of fauna that had populated the entirety of South Asia. Surgeon John G. Malcolmson, who found fossils along the banks of the Narmada River in central India, also believed that these fossils were connected to those found in the sub-Himalayan region.[45] In 1840, Richard Baird Smith, another British engineer who worked as an assistant to Cautley in the Doab project, suggested that the fossils of the Himalayas were similar to those of lower Bengal. He believed that there was "complete analogy" between the fossils of the Yamuna, lower Bengal, and the Narmada and those found along the banks of Irrawaddy in Burma.[46] He found that the "succession of strata" was "perfectly identical" and that they belonged to the "same mighty ocean which reached to the foot of the mountains." The fossilized shells belonged to this ocean and were deposited in different locations because of movements of elevation and depression.[47] This was one of the earliest speculations that an ocean could have preceded the Himalayas.

While these speculations gained force in India, the grand theorization took place in London. In 1838, Cautley decided to dispatch most of the fossils from the Saharanpur museum to London, "the centre of all science," with free access to those "desiring to study and compare" them.[48] From 1840, Falconer and Cautley put the fossils into several boxes and sent them by boat down the Yamuna and then the Ganga to Calcutta. From there, they were shipped to the British Museum in London. Falconer accompanied the last batch of fossils to London in 1842. He remained in England until 1846 and analyzed the entire collection of Indian fossils at the British Museum. In 1845, he sent his report to the Court of Directors.[49] This led to the publication of his *Fauna Antiqua Sivalensis* in 1846.[50]

In London, where he had access to fossils from the various river basins and the Himalayas, Falconer extended these speculations and came up with his grand theory of the formation of ancient alluviums in the river basins of India. He explained the formation of the Siwaliks, the Narmada and Irrawaddy basins, and more significantly, the Indian subcontinent itself. Comparing the *Sivatherium* of the Siwaliks with the *Dinotherium*, giraffe, mastodon, elephant, rhinoceros, and hippopotamus of Cambay, and with the fossil shells of the Irrawaddy, he asserted, "The mass of the Perim fossil belongs to the same genera and species which are found in the Sewalik hills, and in the ossiferous beds of the Irawaddi in Ava."[51]

Falconer produced a large color map to elucidate his theory (fig. 2.2). He suggested that at an "early Tertiary period,"[52] the Indian subcontinent was a large "ancient island" separated by a stretch of water from the Himalayas and the Hindu Kush mountains. The present valleys of the Ganga and the Indus were essentially long estuaries into which the Himalayas deposited their silt and alluvium. They are shaded on the map. An upheaval led to the rapid deposition of silt and converted these straits into the plains of India, connecting

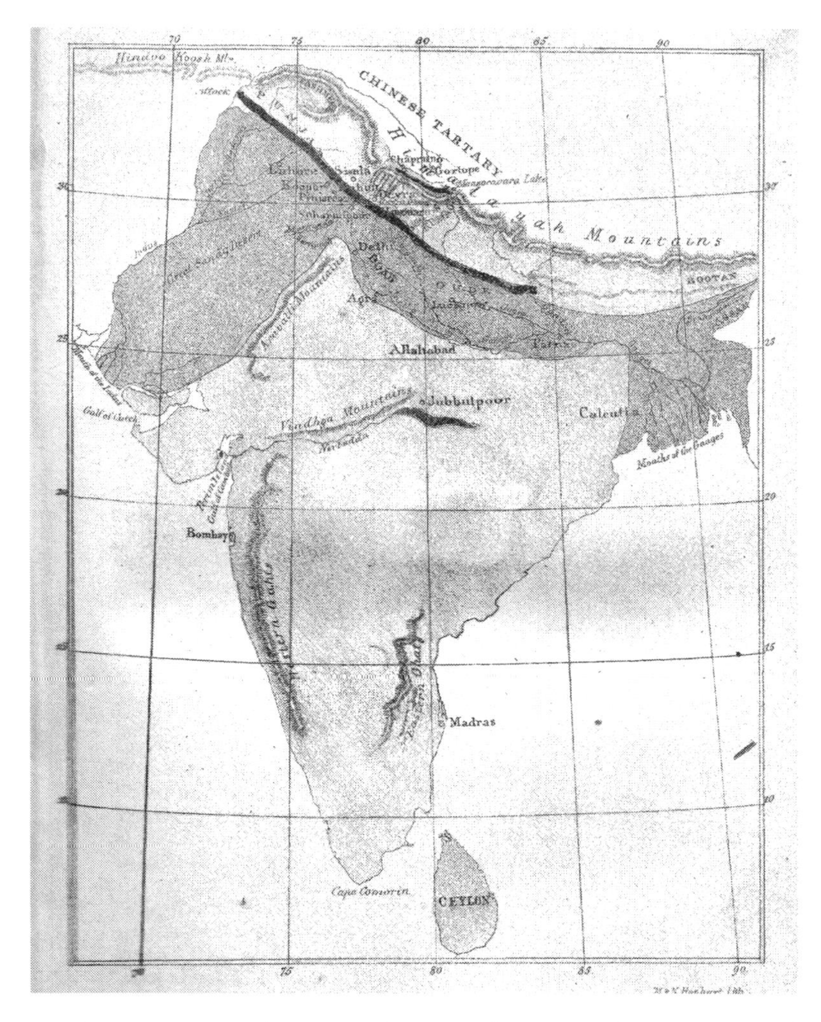

Fig. 2.2. Falconer's map of the ancient alluviums of India. Charles Murchison, ed., *Palaeontological Memoirs and Notes of the Late Hugh Falconer* (London: Robert Hardwicke, 1868), 1:6, plate 2.

them with the island and forming the present subcontinent. The Siwalik fauna then spread all over the Indian subcontinent along with the alluvial deposits (shaded), from the mouths of the Irrawaddy to the Gulf of Cambay. Once the fauna populated the region, a subsequent and even greater upheaval elevated the narrow belt of the Indo-Gangetic Plain into the Siwalik ranges. This and subsequent changes in climate led to the extinction of the Siwalik fauna, although similar fauna survived in the Gangetic Plain of northern India.[53] He believed, as he once wrote to Lyell, that the central Indian fossils found around the Narmada River were of a later age than the Himalayan ones, as they evolved from the primordial Siwalik flora and fauna.[54]

This was the first exposition of Falconer's theory of "ancient alluviums," the soil common to all the riverine basins of the Indian subcontinent, which contained most of the fossils and thereby the clue to primordial life of this region. Lyell originally used the phrase to denote the landscapes carrying traces of the biblical deluge, drawing from the diluvial theory that was popular in Europe in the 1820s, particularly in the works of William Buckland.[55] In Europe, diluvianism was replaced by theories of glaciation by the 1840s.[56] Falconer referred to this slightly older phrase without its biblical connotations to provide a new explanation of deep time in India. In subsequent years, as we shall see, he used "ancient alluviums" even more widely, as a natural-historical metaphor for the birth of civilization in India.

This theory of the creation of the Indian subcontinent was proposed decades before the continental drift theory. Falconer did not, therefore, suggest that the subcontinent had moved and pushed into the Asiatic mainland, as Alfred L. Wegener and Alexander L. du Toit would almost 90 years later; this idea was not accepted universally by geologists until the mid-twentieth century. Rather, he proposed that the alluviums of the river valleys led to the formation of the Himalayas and sustained the flora and fauna of the Indian subcontinent. Therefore, the ancient alluviums of India and its fossils were crucial to understanding the emergence of life in India. To Falconer and other geologists in India, there was another secret buried in these ancient alluviums: the origin of humans, to which I will turn next. In subsequent years, this predominately geological account of the evolution of the subcontinent evolved into a more holistic theory about the deep past of India that incorporated the Orientalist encounters with Indian antiquity in the nineteenth century.

Falconer's theory of the formation of the Himalayas and the Gangetic Plain had practical engineering applications as well. In subsequent years, Cautley debated another British engineer, Arthur Cotton, about his modes of constructing the Yamuna and Ganges Canals. Cautley had to defend his decision to construct the cut of the Ganges Canal above the commencement of the highland of the northern Doab where the river runs upon shingles on a high incline. Cotton's position was that it should have been below that line and that the water should have been channelized by a dam, as he had done on the River Godavari in the south when he built the famous Kallanai Dam.[57] In his official response, Cautley explained his reasons for not constructing a dam by referring to the theory of the formation of the Himalayas. He clarified that the Siwaliks were constituted of the same soil as the existing beds of the Ganga and the Yamuna, only in an elevated state, "unconsolidated beds of sand, clays, and boulder-gravel, or of loosely consolidated and crumbling sandstones." The porous soil and brittle rocks, formed mostly of the erstwhile alluvium of the river basins, were unsuitable for the construction of dams, unlike the much older rocks in the Godavari basin that Cotton had used.[58] He pointed out that Mughal architects had likewise avoided using the Siwalik stones for building purposes and instead brought in sandstone from other places.[59]

Orientalism and Prehistory: Primeval Humans in the Ancient Alluviums

Questions of human origin in the Indian subcontinent had featured throughout the discovery of the Himalayan and Yamuna fossils. In 1836, Baker and H. M. Durand, two engineers employed in the Doab Canal project, found the fossil remains of a primate in the Siwaliks near the Sutlej River.[60] They believed that the specimens belonged to a "gigantic" species of ape contemporary with other fossils found in the region. This discovery, according to them, addressed the "desideratum in Palæontology—[the] proof of existence, in a fossil state, of the type of organization mostly nearly resembling that of man."[61] A few months later, Falconer and Cautley found the fossil teeth and jaw of a primate in the Siwaliks as well. They too believed that these were the fossils of an animal "so nearly approaching man."[62] They are in fact, believed to be the first discoveries of fossil primates.[63] They also initiated one of the earliest paleontological deliberations on the links between primates and humans, decades before the Darwinian propositions and the debates around the Neanderthal in Europe.[64]

Falconer and Cautley wrote a paper jointly in 1836, which was eventually published in the *Transactions of the Geological Society of London* in 1840.[65] Here they referred to more primate bones they had found, including facial and leg bones. They stressed that the fossils perfectly resembled living animals, which suggested a peculiar continuity between extinct and existing species in India, a "mixture of the new and of the old, of the past and of the present, of familiar with surprising forms, together with a numerical richness, such as no other explored region has exhibited within so comparatively limited a space."[66]

They came to this conclusion partly because of their method of work in the remote Himalayas. As they did not have access to other fossils to compare at Saharanpur, they collected and slaughtered animals such as buffaloes, antelopes, other quadruped mammals, and reptiles from the neighboring forests, hills, and riversides and compared their skeletons with those of the fossils.[67] This allowed them to find a "striking" similarity between the existing and the extinct fauna of India.[68] Falconer suggested that there was "one comprehensive fauna" in India and the fossilized species were very similar, if not identical, to the living ones.[69]

This idea of an unchanging Indian natural world, of species lost from the evolutionary chain, also resonated with the general nineteenth-century European perceptions of Indian nature, a changeless entity where time stood still. In the deep forests of the "Terai" in the foothills of the Himalayas, the ornithologist Samuel Tickell had found that "the light of day seldom enters, and the cadaverous weeds, fixed in a stagnant atmosphere, never wave in the refreshing breeze;—afford asylums to the rarer and wilder animals of the forests." He was convinced that the indigenous people there had seen the hippopotamus, which could be found only in Africa. He once caught, in the thickly forested tribal belt of the valley of the Subarnarekha River in southwestern Bengal, a glimpse of an animal he believed was an orangutan.[70]

The Himalayan fossils, therefore, were vital to Falconer's proposition of the faunal continuity in the Indian subcontinent, which enabled him to suggest that it enjoyed relatively unchanged ecological and geological conditions for faunal life. It is possible to see in his theory of ecological and faunal changelessness aspects of Cuvier's ideas of fixity of species. The mummies brought into France following Napoleon Bonaparte's invasion of Egypt in 1798 enabled Cuvier to compare living and mummified animals. Based on these and other observations, he suggested that there was no measurable difference between them and argued that there was fixity of species that each species constituted an ideal form that cannot change over time. However, Falconer's emphasis was on the unchanging ecology, the "constancy in the order of nature, of an identity of condition in the earth of the olden time with what it exhibits now," while Cuvier's stress was on species.[71] Moreover, Falconer's work belonged to the age of prehistory, he referred to prehistoric fossils for comparison, and he was, as we shall soon see, concerned with prehistoric humans. Cuvier in his lifetime remained unconvinced of prehistory in general and was vague about the prehistoric origin of humans in particular.[72]

The distinctiveness of Falconer's paleontological research was in its composite understanding of Indian antiquity derived both from prehistory and Orientalism. Since the end of the eighteenth century, the Asiatic Society, established by William Jones in Calcutta (1784), had studied the antiquity of India in its natural and cultural worlds. It had investigated classical texts, inscriptions, and rituals, along with the plants, rocks, climatic patterns, and natural phenomena. In the words of its founder, in his inaugural speech to the society, "If now it be asked, what are the intended objects of our inquiries, within those spacious limits, we answer, MAN and NATURE, whatever is performed by the one, or produced by the other."[73] "Man" here represented the Oriental self, its culture, language, tradition, and myths; and "nature" represented Asia's physical world and natural history. The society and its members thereby established some of the basic features of the conjoined studies of the cultural and natural landscapes of the Indian subcontinent that I am analyzing here. Jones himself studied Indian botany and its Sanskrit nomenclature, which he compared with Linnaean nomenclature and classification, alongside his studies of Indian languages, legal systems, religions, and the Puranas.[74]

In 1823, the Sanskrit scholar Henry Thomas Colebrooke established the Royal Asiatic Society of Great Britain and Ireland in London, a rare occasion when a metropolitan institution succeeded its colonial counterpart. The Royal Asiatic Society became one of the main centers for the cultural, historical, and natural-historical study of Asia in Britain in the nineteenth century. In 1830, the Literary Society of Bombay (1804) was affiliated with it and became the Bombay Branch of the Royal Asiatic Society. The same year, the Madras Literary Society (1817) became associated with it as well. Before the formation of specialized Indian surveys such as the Geological Survey of India and the Archaeological Survey of India, they were the only places for the study of and deliberations on Indian geology, archaeology, and ethnology, alongside various cultural and historical investigations. These close associations between the natural sciences and cultural studies in the early nineteenth century formed the basis for Falconer's subsequent paleontological discussions, and his understanding of Asiatic culture shaped his ideas of the prehistoric Asiatic "man."

In London, where he had arrived with the Siwalik fossils in 1842 and where he closely studied the various fossils of the subcontinent, Falconer became a regular visitor to the Royal Asiatic Society. He gave two lectures at two meetings there in 1844, where he started by paying tribute to the philological, ethnological, and cultural research of the society. He then connected its research with his recent paleontological investigations of Indian fossils. He drew parallels between Indian cultural antiquity and the paleontological deliberations about the origin of the human race in India. He declared that a civilization as ancient as India, discovered by the Asiatic Society, must have also been the cradle of prehistoric humanity; "a large section of mankind first dawned in the valley of the Ganges."

> The antiquities and literature of the East have been, from its commencement, the special field of investigation to this Society, and to the parent institution in Calcutta. A rich vein has been opened, branching in a thousand ramifications, and fertile in results of the deepest interest. The human race has been traced farther back into time in the East than in any other quarter of the globe; and the tendency of all inquiries has been to show that the civilization of at least a large section of mankind first dawned in the valley of the Ganges.[75]

In this talk, Falconer introduced the Royal Asiatic Society to the theme and the new possibilities of prehistory that had captivated the contemporary scientific community of Europe. Although since the late eighteenth century members of the Asiatic Society had studied the geological and paleontological features of the Indian subcontinent and Asia, there had been little discussion on the meanings and implications of deep time and prehistory. Falconer was in some ways presenting the thesis that Prinsep had asked for a few years back in his statement following Tournal's propositions and the discovery of the Yamuna fossils (see chapter 1). He suggested that fossil research shared the intellectual inspirations and methodologies of the Asiatic Society. The connection he made between antiquarianism and paleontology was different from the eighteenth-century propositions of Blumenbach, who had placed fossil research alongside general antiquarian studies.[76] Falconer's work, on the contrary, belonged to the era of prehistory, a period when the lines between the "geological" and the "historical" were drawn. Falconer needed to establish fresh links between prehistory and nineteenth-century antiquarianism.

He first commented on the temporal limits of conventional antiquarian studies: "There is a point up to which we can follow man back through the records of language and art and the shadowy indications of mythology and tradition, but beyond which we cannot go. Every trace of the human race then fails us."[77] The need was to adopt a new method to extend the study of human antiquity deeper into the past: "If we desire to dive further into antiquity, we have to fall back on the monuments and inscriptions constructed by nature, on the fossil remains of the extinct races of animals which formerly peopled the earth."[78] He then provided a detailed account of the discovery of fossils in the Siwaliks and the deep history of the river basins of the Yamuna, Irrawaddy, and Narmada. He explained his thesis of the "ancient alluviums" referred to earlier, the formation of the Himalayas and the Indo-Gangetic Plain, and faunal migration in the Indian subcontinent. He suggested that India, more specifically the Gangetic Plain, with its relatively constant ecology had the ideal conditions for human evolution:

> Man, *cæteris paribus*, must have progressed most rapidly where most favourably placed in regard to the external conditions which regulate the increase of his race and the development of his social relations. Neither the valleys of the Nile, nor of the Euphrates, Tigris, or Oxus, in extent and fertility together, or in the richness and variety of their productions, can admit of comparison with the valley of the Ganges.[79]

Falconer made an intriguing suggestion here about early humans and the favorable conditions of their emergence in the river basins of India: "To benefit suitably from such favoured circumstances, we find that the Indian variety of the Caucasian branch of the human family has the most perfect development of that physical conformation which is observed to be associated with the highest capability for mental improvement."[80] The phrase "Indian variety of the Caucasian branch of the human family" highlights the mixed references he frequently made about early humans in this talk and the blend of Orientalism and prehistory that marked his work in general. Here he drew from a long tradition of religious and philosophical studies in Britain, which also featured in Anglophone Orientalism, that saw Hinduism, and in particular Brahmanism, as part of a shared cultural heritage with Europe. These links were often articulated through ideas of Druidism, which contained nascent ideas of Aryanism. Druidism was crucial to the understandings of primordial religion in Britain and the construction of British self-identity from the late eighteenth century. Philip Almond describes the period between 1740 and 1840 as the "hey-day of the Druids" in Britain, as druids featured in various public debates and displays.[81] India was a significant point of reference in these deliberations, as Druidism was often believed to have originated in the East. The Scottish philosophers and writers Thomas Brown (1778–1820) and James Macpherson (1736–96) speculated about the eastern origin of Druidism and associated it with the Brahmanical traditions.[82] The English religious scholar Godfrey Higgins (1772–1833) also believed that Druids were Oriental priests who had migrated to Europe from India.[83] The Orientalist scholar Francis Wilford also indicated strong links between Hinduism and the primordial religions among the Greeks, Romans, and Egyptians. This was part of his belief that ancient civilizations shared the universal religious lineage, reflecting the same faith in the Supreme Being.[84] The Orientalist mathematician Reuben Burrow (1747–92) claimed that the Druids were Brahmins who migrated to the colder climates of Europe.[85] Thomas Maurice (keeper of Oriental manuscripts at the British Museum in London), in his *Indian Antiquities*, published in 1796, also claimed that Druidism came to Europe from Asia along with a group of priests with the Japhetic tribes, professing the Brahmanical religion. They also entered northern India and formed the "Indian Caucasus" race.[86] Therefore, the contemplation of the Brahmanical connections to Druidism was part of an early nineteenth-century Orientalism and Aryanism in which India featured centrally. Although Falconer did not directly refer to Druidism, he invoked ideas of the primordial origin of humanity in India that verged on the Asiatic Druidism of Burrow or Maurice to suggest that primeval humans originated in India. He proposed, drawing from the discovery of the Himalayan fossils and the ensuing Orientalist discussions about primordial humans, that the Gangetic alluvial plain, because of its peculiar geology and ecology, was the site of the origin of these primordial races.

These suggestions were still conjectural. Falconer merely presented his geological theory of the formation of the Indian subcontinent to an Orientalist audience and mixed it with references to the studies of Indian antiquity. His hypothesis of India as the cradle of human civilization crystallized into a theory when he returned to the subcontinent. In 1845, having completed his survey of the riverine fossils of India in Britain, he came back to India. This time he was based not in the Himalayan foothills but in Calcutta, which was one of the centers of nineteenth-century Orientalist scholarship. He was appointed professor of botany at the Medical College in Calcutta and then in 1848 superintendent of the Calcutta Botanical Garden. He was also a regular visitor to the Asiatic Society of Bengal based in that city. While he was away from the main collection of Indian fossils, which was now in London, other intellectual encounters changed the course of his investigations. In Calcutta, Falconer was deeply influenced by the Indo-centrism that was fundamental to Orientalist scholarship around the middle of the nineteenth century. His thesis on the geological evolution of India, in turn, became part of a grander and more enduring idea of India as the cradle of human civilization.

Falconer did not conduct any new geological explorations during his second visit to India. He worked on the remaining fossils in the Asiatic Society's museum in Calcutta and on some of the old Yamuna fossils. At the request of the Asiatic Society of Bengal, he prepared a catalog of the remaining fossils in its collection with the help of Henry Walker, professor of anatomy and physiology at the Medical College in Calcutta. The catalog was published in 1859.[87] Falconer found the task challenging. Fossils from the Siwaliks, mostly without labels, were mixed with those from Perim, Narmada, and Ava. Often, the provenance he assigned to them was, he admitted, "simply conjectural."[88] He depended on his familiarity with these fossils, their distinctive colorations, and their anecdotal lineages to order them in a new classification.

Almost forty years later, in 1885, the English geologist Richard Lydekker faced a familiar challenge in cataloging the Siwalik fossils of the Indian Museum in Calcutta. As he worked through the collection, Lydekker realized that it included specimens gathered at different times and from different parts of India. Moreover, their exact provenance and lineage could not be authenticated. Like Falconer, he found that the "large" Siwalik set was placed along with those of Perim and Burma, as well as the Narmada fossils, when they were moved from the Asiatic Society of Bengal to the Indian Museum in 1875. Many were "generically undeterminable."[89] Various individuals had donated several pieces of fossils to the society, while a small series of specimens from other sites included items acquired by the Indian Museum before 1875 or those "transferred from the Asiatic Society of Bengal without history."[90] Lydekker identified specimens from Sindh, Kutch, Siwalik, Burma, and Perim island.[91] The Yamuna fossils "appear[ed] equivalent to those of the Narbada," while the fossils from the River Painganga in western India appeared similar to those from the Narmada.[92]

These experiences of classifying fossils of the subcontinent are important. On the one hand, they reveal the methodology that geologists in India often followed, which was based more on theoretical conjecture than on detailed empirical examination and comparison of fossils. This allowed Indian fossil research to remain an inexact science, borrowing creatively from Indology, zoology, archaeology, and philology throughout the nineteenth century. On the other, there is certain appropriateness in this chaotic amalgamation of fossils and their lineages. This lack of order, discipline, and fixity of the fossils of the Asiatic Society of Bengal and the Indian Museum reflected the hybrid and concurrent intellectual experiences through which they were discovered and theorized. To order his fossils in Calcutta, Falconer used contemporary Indological philological deliberations of Indian antiquity as evidence. At a time when European geology was rapidly acquiring distinct methodologies and fieldwork techniques, in Calcutta we find an intellectual tradition that remained resolutely interdisciplinary.

The work on the catalog allowed Falconer to revisit the theory of "the 'Ancient Alluvium'" and its possible bearings on "the antiquity of man."[93] He now believed that the Yamuna fossils of Edward Smith and Dean were "the most promising of results bearing upon the human period."[94] His theory now developed in deliberation with scholars associated with the Asiatic Society of Bengal. During his stay in Calcutta, Falconer interacted with Sanskrit scholars such as Radha Kanta Deb, who at that time was compiling the encyclopedic dictionary of Sanskrit words, *Shabda-kalpadrum* (1857).[95] Deb introduced him to the world of Sanskrit scholarship, comparative philology, and Puranic traditions. Deb also carried within him a deep strain of Hindu civilizational essentialism and Indo-Aryanism. In the 1840s, British scholars introduced Aryan race theories to India.[96] This generated a complex response among Hindu scholars such as Deb who linked Hinduism with Aryanism. Deb rejected the European Aryan race theory, which suggested that Aryans came to India from central Europe and instead argued that India was the original home of Aryans and modern European languages were offshoots of the ancient language of India.[97] Around the same time, Friedrich Max Müller proposed, as a response to this Indo-centric Aryanism, what is referred to as the "two-race theory of Indian civilization."[98] He first put this thesis forward in a paper he presented at the meeting of the British Association for the Advancement of Science in Oxford in 1847, "On the relation of the Bengali to the Arian and aboriginal languages of India." He suggested the existence of two essential races in India, the indigenous races of darker complexion and fairer-skinned Aryan immigrants. Following the Aryan conquest, large parts of the indigenous population lost their indigenous languages and cultures and adopted those of their masters. Others in more remote areas retained their aboriginal cultures. These propositions were based entirely on philological deliberations without any anthropological or paleontological references.[99]

The Indo-centric Aryanism of Deb and the two-race theory of Max Müller infected Falconer's paleontological speculations about human antiquity. Primarily through his work, Indian fossil research was subsumed into contemporary debates on the origins of the Aryans. Although unlike Deb, he believed more in Müller's scheme that Aryans were "emigrants" to the subcontinent and that there were indigenous races there before them, he also suggested, drawing from his previous addresses at the Royal Asiatic Society that the fertile Gangetic Plain was the original home of humankind. Falconer's fossil research became linked to Deb's philological project when he searched for references to the hippopotamus in Indian traditions. As Falconer compiled his catalog in Calcutta, the fossil of a hippopotamus that he and Cautley had found in the Siwaliks in the 1830s posed a problem. The living hippopotamus could be found only in Africa. Philology provided a possible explanation. Falconer consulted Deb and works of other Sanskrit scholars such as Wilson and Colebrooke and learned that there were several words for the hippopotamus in Sanskrit, although the animal did not exist in the subcontinent in living or historical memory. This intrigued him. He joined paleontology and philology and wondered whether the Sanskrit references to the hippopotamus and the fossils of the same animal had a common origin: "May not this extinct Hippopotamus have been a contemporary of Man?"[100] He wondered whether the Aryans, who were believed to have composed the Sanskrit texts, ever saw the hippopotamus "living on the northern rivers of India."[101] His conclusions were that the Indian hippopotamus had gone extinct long before the arrival of the Aryans. However, "it was familiar to the earlier indigenous races," which is how the Aryans learned about it. "It is therefore in the highest degree probable, that the ancient inhabitants of India were familiar with the Hippopotamus as a living animal."[102] This secret was buried "deep in the alluvium of the Jumna, or in ancient deposits in the valley of the Nerbudda" in the form of human and animal fossils.[103] While in the earlier discovery of fossils in the Doab, antiquarian pursuits turned into a geological pursuit, in Falconer's work, Orientalist discussions were grafted onto the emerging geological landscape of India.

For Falconer, the suggestion of prehistoric humans in the Indo-Gangetic Plain opened up new possibilities in Orientalist discussions on the origin of races. They were now faced with geological prehistory, a project that in his own words, had both "paleontological and ethnological bearings."[104] Falconer wrote two papers (both published posthumously in 1865) on the questions of human origin, "Primeval Man, and his Contemporaries" and the "Occurrence of Human Bones in the ancient Fluviatile Deposits of the Nile and Ganges." In these papers, he synthesized his lifetime of research to map out the climatic and geological conditions that had led to the emergence of humans. This was part of a larger work on human antiquity that he had started but did not complete.[105] While making these arguments, he looked back at his early Himalayan expeditions with Cautley in the 1830s very differently. He now suggested that they were in search of human origins: "Captain Cautley and myself were constantly on the look-out for the turning up, in some shape or other, of evidences of Man out of the strata of the Sewalik Hills."[106] He also suggested that their Siwalik discovery of primate fossils that resembled the living monkeys implied that these similarities had bearing on the evolution of early humans in the subcontinent.[107] This ecological constancy meant that the region had the same ideal conditions of human habitation from the remotest time: "Every condition was suited to the requirement of man."[108] If all other fossilized animals could still be found living in some form in this region, "why then, in the light of a natural inquiry, might not the human race have made its appearance at that time in the same region?"[109]

He made a further point in his paper on primeval humans that the river basins of the Indian subcontinent, as well as those of Egypt, were the ideal ecological sites of human origin. He posited the Indo-African riverine ecologies against the harsh conditions of Europe, particularly in prehistoric times. The "tranquil succession of deposits" of the Ganga, the Irrawaddy, or the Nile, rather than the "hard conditions of the Glacial period" in Europe, provided "exceptional conditions" for the primeval human:

> Like the Esquimaux, the Tchuktshes, and the Samoyedes on the shores of the Icy Sea at the present day, man must have been then and there an emigrant, placed under circumstances of vigorous and uncertain existence, unfavourable to the struggle of life and to the maintenance and spread of the species. It is rather in the great alluvial valleys of tropical or sub-tropical rivers like the Ganges, the Irrawaddi, and the Nile, where we may expect to detect the vestiges of his earliest abode.[110]

Falconer referred to the discoveries of human fossils in the Nile basin and compared the alluviums of the two valleys. He depicted the two regions as ecological havens for human evolution: "It is there where the necessaries of life are produced by nature in the greatest variety and profusion, and obtained with the smallest effort; there, where climate exacts the least protection against the vicissitudes of the weather; and there, where the lower animals which approach nearest to man now exist, and where their fossil remains turn up in the greatest variety and abundance." By contrast, "the earliest date to which man has as yet been traced back in Europe is probably but as yesterday, in comparison with the epoch at which he made his appearance in more favoured regions."[111]

Now "ancient alluviums" assumed even greater geographical proportions as they became the category for situating the ancient Indo-African river basins as sites of human evolution. This category proposed a fundamental relationship between landscape and human history in India and Africa. Future explorations of human antiquity, Falconer suggested, thus needed to take place in the tropics at the sites of their old river systems, rather than in the caves and strata of Europe. These explorations would also need a combination of geological research, textual analysis, and ethnological investigations.[112]

Therefore, Falconer's primeval man remained one of mixed heritage, a combination of Orientalism and prehistory. The primeval human that he depicted had indefinite racial and paleontological characteristics. Falconer was working with very limited human fossil specimens of India, and he had no direct research experience in Egypt. This is not to suggest that he was unfamiliar with human fossils and the contemporary paleontological debates on human antiquity. After he left India in 1855, Falconer joined Prestwich and William Pengelly in their search for flint implements and primate fossils in Brixham Cave in 1858–59. They sought to link the flint stone implements of the caves to human evolution in the Pliocene era.[113] The report that Falconer wrote jointly with Andrew Ramsay and Pengelly in 1858 based on these investigations had very conventional paleontological references and tone, with no broad thesis about the origin of early humans.[114] In contrast, when he wrote about South Asia or the Nile, he appeared more assured about the deep origin of humans and drew more freely from cultural insights. His familiarity with contemporary Orientalism provided him with greater conceptual scope and maneuverability. In these respects, his work marked the concurrence of nineteenth-century discussions on the cultural and natural landscapes of India.

Falconer's ideas of human origin in Asia and Africa could be seen as a continuity of eighteenth-century ideas of eastern origins of civilized races, which led German Romanticists such as Johann Gottfried Herder (1744–1803) to look to the east, specifically the Himalayas, in seeking to trace the original home of the Aryans.[115] For these similarities with the eighteenth-century traditions, Falconer may appear to belong to what Thomas Trautmann calls the "lost generation of evolutionism between the eighteenth and late-nineteenth century." Trautmann here refers specifically to James Cowles Prichard, who suggested a common origin of humanity in the early nineteenth century. Trautmann argues that later in the century in the face of the new wave of Darwinian evolutionism, Prichard's work fell into "deep obscurity" and became a "blind spot" of evolutionary theory.[116] Shruti Kapila has shown that Prichard's work straddled the metamorphosis of early nineteenth-century Orientalist discourse into late nineteenth-century Indology, between Jones/Colebrooke on the one hand and Max Müller on the other.[117]

Falconer's position was different from Prichard's. He established the ecology of the fertile river basins of the Ganga and the Nile as the essential element in the emergence of early humans, not necessarily in their racial "negritude," which Prichard believed was fundamental to the primordial human race. Falconer was entirely silent about the racial characteristics of primeval humans. Here he reflected an important feature of the search for the primitive humans in the early Orientalist tradition: its indefinite physical features, in contrast to the racialized descriptions in the contemporary anthropology of aboriginal or primitive humans.

More importantly, Falconer's work on the origin of humans in the alluvial plains came almost two decades after Prichard's, right at the onset of the Darwinian evolutionary theories within which Prichard's ideas lost much of their relevance. Rather than being subsumed by Darwinian evolutionism, Falconer's ideas of the origin of humanity in India and Africa gained greater resonance across a wide range of intellectual traditions in the late nineteenth century.

Therefore, it is more useful to see Falconer's later work in reference to subsequent geohistorical explorations in Asia and Africa in search of human prehistory than as a legacy of an older tradition. Falconer died in 1865, leaving his suggestions about the Indo-African origin of humans tantalizingly poised. In the late nineteenth and early twentieth centuries, several geologists took up different aspects of his work on the Siwalik fossils, primates, the formation of the Gangetic Plain and the Himalayas, and the emergence of humans in India and Africa. For example, Prestwich corroborated Falconer's proposal that "civilised man" first appeared in the eastern parts of the world in the Neolithic age. This, he explained, was essentially due to climatic reasons, as Egypt and southern Asia had conditions more favorable to human life at the recession of the glacial age.[118]

Falconer's propositions about the origin of the human race in the river basins of India and Egypt were a precursor to twentieth-century theories of origin of humanity in Africa. He was one of the first to relocate that paleontological question away from Europe. In doing so, he also incorporated the ecology of these regions as key themes in these deliberations. These later investigations repeated a key element of Falconer's work, situating discussions of human antiquity in the tropical and subtropical parts of the world.

A Sunken Continent and a Buried Ocean: The Lost Links of the Deep Past

Falconer's propositions about the Indo-African origins of humanity and the significance of Himalayan fossils to the evolution of faunal life in the Indian subcontinent became key themes of late nineteenth- and twentieth-century explorations of the deep past. They unfolded within both a wider and a more "naturalized" landscape. Ecology and geology became the key determinants in the late nineteenth-century search for human origins. In such discussions, the erstwhile Orientalist or cultural legacies were increasingly overlooked. Yet the imaginations of Indo-African connections of human antiquity first appeared in eighteenth-century biblical readings of history and remained extant within Indological deliberations in the nineteenth century. Jones suggested ancient links between Egypt and the Indian subcontinent, as he believed the Hamites migrated from the former to populate the latter.[119] These biblical themes also led to geological expeditions. Arthur Bedford Orlebar, a college professor and government astronomer in Bombay, theorized about the existence of an antediluvian ocean "long anterior to Mosaic creation."[120] As a devoted evangelical Christian, he searched for geological traces of biblical history in Asia and Africa. Following his visit to the Mokattam hills, south of Cairo, he became convinced that "our geological facts then perfectly agree with the Bible history; for it is certain that a great flood was the last geological event in Egypt, and this affords additional evidence to the truth of the Bible record."[121] This biblical motif of African migration to India and the Indian Ocean region resonated in ethnological and anthropological studies throughout the nineteenth century. Africa, particularly its ancient civilizational connections with the Orient, featured in Orientalist literature during this time. Horace Hayman Wilson, the long-term director of the Royal Asiatic Society in London (1837–60), declared in his lecture to the society in London in 1852, "The East, however, is a relative term." He suggested that the society's study of human and cultural antiquity of the East include Asia, sub-Saharan Africa, and particularly Egypt. While the "antiquities of Egypt have an Orientalism of their own," its hieroglyphics, along with those of Syria, formed an integral part of Orientalist studies.[122]

These predominantly ecclesiastical and Indological deliberations were redefined by geohistorical discussions. Zoologists and geologists noticed similarities between African and Indian fauna and in the geological formations of the two regions. One significant difference between Falconer and others was that while Falconer had suggested the ecological similarities of the two river basins, later scholars proposed the existence of an actual geological link between them. In 1867, the British Indian geologist W. T. Blanford traveled to east Africa from Bombay as part of the British Abyssinian expedition. The expedition was composed almost entirely of troops and officers sent from India. As the British troops besieged and destroyed the fortress of Magdala, Blanford explored the antiquity of the surrounding landscape. He noticed that the fauna of a large portion of the Indian peninsula had a strong affinity with African fauna.[123] He also found that fossils of hippopotamuses, camels, rhinoceros, and antelopes found in the Siwaliks showed strong affinities with the African fossils.[124] Apart from fauna, Blanford also found the geological features, such as the Deccan traps in central and western India, to be similar to the trap formations of Abyssinia.[125] Based on these ideas, his brother, Henry F. Blanford, suggested in 1875 that a landmass existed between the two regions.[126] Around the same time, another English zoologist, Philip Sclater observed that the mammals of Madagascar shared characteristics with those in continental Africa and Asia, which he explained through the presence of a great continent connecting Asia and Africa, most of which was at present submerged under the Indian Ocean. He named this lost continent "Lemuria."[127] As Ramaswamy has shown, to Sclater and other British naturalists, Lemuria, and the Indo-African connections that it represented, was the lost paradise, a primeval land that held secrets of the "cradle of mankind."[128]

Suess adopted these ideas to suggest the existence of an "Indo-African table-land." This formed the key component of his proposition of the presence of a prehistoric Gondwanaland, which he believed was formed of several such land bridges across the oceans.[129] Although Falconer did not actively participate in all these debates, some of which developed after his death, his ideas were absorbed into these emerging debates about the prehistoric connections between India and Africa.

In the 1870s, Lydekker came across several fossils belonging to primates in the Siwaliks, in the same region where Falconer and Cautley had found theirs four decades earlier. William Theobald, a geologist belonging to the GSI, sent Lydekker a new primate fossil that was found in the Siwaliks that did not conform to any of the known ones. Lydekker named it *M. sivalensis*.[130] He later came across another specimen found by Theobald in the Punjab part of the Siwaliks, which he named *Palaeopithecus sivalensis*. These were the first evidence, according to Lydekker (apart from Falconer's discovery in the 1830s), of the presence of "a large anthropoid ape" in India. He connected these fossils with discoveries of similar anthropoid apes in Africa and Sumatra to suggest a "common" parentage and "ancestral home" of hominoids in the "sunken southern continent" of "Lemuria" or "Indo-Oceania."[131] He believed that the secrets of human evolution were buried in that continent under the Indian Ocean and that "we may never discover the 'missing links.'"[132] The only hope was to search in these tropical countries, the coastlines of the sunken continent, for clues of the lost land and of the lost humanity that bridge the gap between primates and humans.[133] While the Indo-African continent and its promised human links remained hidden, the Indian subcontinent and the eastern parts of Africa became the main sites of paleontological investigations into Indo-African faunal and human antiquity.

Himalayan fossils also suggested a new thesis of human origin in the late nineteenth century. In 1893, Suess gave the name "Tethys" to the great ocean that he believed once stretched across Eurasia, the remains of which could be found in the lakes of Tibet, the Alps, and the Mediterranean Sea.[134] Geologists believed that the Tibetan lakes and the Karewas (dry lakes rich with marine fossils) of Kashmir were the last remains of the Tethys.[135] The Himalayan upheaval, which was believed to be the most recent geological event in this vast Eurasian region, raised the beds of this ancient ocean, and therefore Himalayan fossils carried traces of the history of Tethys and even of humans, who they believed appeared first along the shores of the prehistoric sea. According to geologists, the Himalayas rose over three distinct phases, starting in the Eocene period; the last phase was in the Pliocene era, which was linked to the evolution of humans. Now the earlier discoveries of primate fossils in the Himalayas acquired greater significance.

In the early twentieth century, following these propositions, geologists took a renewed interest in Himalayan fossils, particularly because they appeared to hold the secrets of human origin. The American geologist Joseph Barrell and the British paleontologist Arthur Smith Woodward made the first connections between the rise of the Himalayas and the evolution of humans, which incorporated Falconer's theory of climatic changes and faunal evolution in the subcontinent. Barrell and Woodward reflected on the theories of the rise of Himalayas and the discoveries of primate fossils to argue that a "varied assemblage of apes" lived in the forests of northern India before the Himalayas existed, and primitive humans evolved from them at the time of the rise of the Himalayas. As the mountains rose, Smith Woodward claimed, the temperatures dropped, and some of the primates living in the warm forests were trapped in the northern raised areas. Those who survived adapted to the new environment and became carnivorous. With the continued development of their brain and physique, they became humans.[136]

Motivated by these suggestions, the German geologist Hellmut (also spelled Helmut) de Terra, based at Yale University, led two expeditions in the Himalayas in the 1930s, along with other Yale geologists and biologists. They discovered several Paleolithic stone flakes in different parts of the Himalayas.[137] At the end of the expeditions, de Terra declared that they found some the earliest traces of prehistoric humans in the Himalayan region.[138] As a member of the expedition, the American paleontologist G. Edward Lewis found several fossils of primates in the Siwaliks, which were subsequently presented to the Yale Peabody Museum of Natural History.[139] These finds initiated a long and protracted debate regarding the various hominid fossils discovered in South Asia—labeled with semi-mythological names such as *Ramapithecus* and *Sivapithecus*—which served as the basis of claims that India was the cradle of human civilization.[140]

The British geologist Guy Ellcock Pilgrim, who acted as an advisor to de Terra during his Himalayan expeditions, provided the main theoretical framework of this thesis. He came to India in 1902 with formal geological training in Britain and sought to assign clear stratigraphic and chronological order to the Siwalik fossils.[141] In the process, he also attempted to align these fossils more closely with European geological periodization.[142] Despite this initial Eurocentric ambition, he soon proposed an eclectic theory of river basin formation in northern India (based on his study of the fossil vertebrates of the Siwaliks) that closely corresponded with Falconer's theories. He believed that the primate fossils of the Siwaliks he had found and named *Sivapithecus indicus* were those of "the direct ancestor of man."[143]

The theory of the relatively recent formation of the Himalayas became the mainstay of the suggestions of primeval origin of humans in the Indian subcontinent and interestingly marked a return to the cultural and natural landscape of the Doab and the Indus. Birbal Sahni and Wadia were the two most prominent Indian geologists working on Himalayan fossils in the early twentieth century. They suggested the rivers of northern India as the site for the emergence of primitive humans—not just the Siwaliks and Tibet—before the rise of the Himalayas. They merged the nascent theories of plate tectonics with that of the existence of the Tethys, as well as the theories of human evolution in the Himalayas, to suggest a new ecology of human evolution. According to Sahni, the Himalayas were once the site of the Tethys Sea, which was like a "Mediterranean ocean" and the habitat of primitive humans. These populations either perished or were separated from their Central Asian counterparts as they migrated south following the rise of the Himalayas. Although the Neolithic humans of India had, therefore, lost their Central Asian lineages, they remembered the Himalayan landscape, the flora, and the fauna. This, according to Sahni, explained why ancient Himalayan medicinal products such as the *silajit* have been used in India since "time immemorial."[144] He even believed that the recently discovered Harappan script found in the Indus Valley in the 1930s, which had mystified contemporary paleographers, bore elements of Chinese script and thus traces of this lost history of humanity.

As he searched the remote and barren Himalayan landscape, Sahni also felt the need for a more denaturalized narrative and a wider intellectual pursuit to connect the various lost and living traces of prehistoric humanity. He urged his colleagues to reconnect paleontology and geology with other antiquarian studies:

> In this age of specialisation, which inevitably tends to confine thought to compartments, one is apt to overlook or underrate the bearings of one branch of science upon another. A palæobotanist or a geologist, accustomed to think of Time in millions of years, stumbles upon an archaeological discovery, which at once brings him down to human epoch. It forces his attention to the wanderings of man since the time he began to leave signs of his handiwork in the form of stone or metal implements, inscriptions, coins, seals, or other monuments of his ever-increasing intelligence and power.[145]

Conclusion

This chapter has traced how the Orientalist discovery of Indian antiquity came to reside within the geological narrative of the "ancient alluvium." This, in turn, provided Orientalist antiquarian studies with a greater naturalistic scope. The theories suggested by Falconer, Lydekker, Pilgrim, Sahni, and Wadia, which linked human antiquity with the geological transformations of the Himalayas and the northern Indian plains, appeared persuasive in their time because they belonged to their contemporary landscapes of antiquity. In its seepage into cultural antiquity, geology underpinned the history of India as much as it was defined by it. By the mid-twentieth century, other landscapes and other river basins, such as the Rift Valley of the Olduvai River in Tanzania, and alternative imaginations of human evolution formed the basis of new theories human origin.[146] The Siwalik fossils were subsequently ensconced within the theory of continental drift, which explained the formation of the Siwalik and Indo-Gangetic landscape through the movement of tectonic plates. Therefore, the episodes discussed here may appear fantastic and even fictional from our contemporary geological and paleontological perspectives. For that very reason, they are vital for us to revisit. They enable us to appreciate that the geohistorical theories of the earth and questions of human origin had, and continue to have, multiple associations and meanings, which underwent significant metamorphoses, despite the progressive and naturalized narrative that geology presents today. The scientific facts about and provenance of these fossils, lying heaped in a creative mess on the shelves of the Indian Museum in Kolkata—as they still lie—were established in reference to Orientalist and Indological intellectual traditions that had ostensibly little to do with paleontology. It was within such an approach, as we shall see in the next chapter, that these fossils were also interpreted through Indian mythological hermeneutics.

Here I have just broached the question of human antiquity in India. As we shall see in chapters 4 and 5, this question became central to Indian geology and ethnology. In this particular instance, the search for the origin of humans was part of a pervasive search in the nineteenth century for the origins and evolution of nature itself—of rivers, seas, mountains, and the soil. At the same time, the alignment of colonial fossil research with colonial antiquarianism allowed unique hypotheses to emerge. The alluvium, the rivers, the mountains, their inhabitants, and their myths made sense within this natural-historical frame of antiquity.

Mythic Pasts and Naturalized Histories

The Deep History of Sacred Geography

Sacred geography emerged as a significant theme in seventeenth-century antiquarian studies in Europe. The impetus came from the resurgence of Mosaic history and theology, which motivated biblical explanations of global geographical and natural experiences of the earth. Robert Fludd's *Philosophia Moysaica* (1638), Joachim Becher's *Physica Subterranea* (1669), Thomas Burnet's *Sacred Theory of the* Earth (1681–89), and Thomas Robinson's *The Anatomy of the Earth* (1694) established Mosaic history as the mainstay of the history of the earth.[1] Sacred or biblical geographical themes emerged as key points of reference in these texts. The classical scholar Jacob Bryant in *A New System; or, An Analysis of Ancient Mythology* drew the template of this grand theological and historical scheme (published in three volumes between 1774 and 1776). He reinstated Mosaic history as universal human history, which he suggested had been distorted by subsequent fables. The book sought to present the "true history" from diverse traditions across the world and restore it to the "original purity" of Mosaic history, which narrated the grand scheme of the deluge and the subsequent distribution of the human race through Noah's three sons. He suggested that references to Noah and the deluge could, therefore, be found in all human civilizations.

Bryant's work became the blueprint for the eighteenth-century search for Mosaic historical traditions in different parts of the world, marking the beginning of European discourses on non-Abrahamic religions, such as Buddhism and Hinduism. From the second half of the eighteenth century, European religious scholars speculated about the antiquity of these religious traditions. Scholars such as William Jones searched Sanskrit texts for biblical references. These endeavors provided Oriental nature with sacred biblical connotations. These biblical themes, in turn, became integral to the Indian deep past.

Deep history, despite its secular geological orientations, remains heavily invested in the mythic past. This chapter investigates how myths, specifically Hindu mythologies, became "geomyths" in the nineteenth century, in other words, how these myths acquired geological undertones. Historians have shown that eighteenth-century deliberations on geohistory were ingrained with biblical myths. The question of the deluge remained important even in nineteenth-century geology, despite the dominant secular motifs of the discipline.[2] Here I consider the other aspect of this history: how mythologies themselves became infused with ideas of deep geological time and nature. This naturalization of sacred geography does not conform to the narrative of the gradual secularization of geohistory. Rather, myths and geohistory came to coexist or became interchangeable categories, which established geomyths as powerful cultural and political motifs in modern India. The problem here is not just in addressing the sacred-secular dyad in the modern understanding of nature but also in tracing the relationship between sacred nature and deep history that was formed in the nineteenth century and continues to this day.

There are two premises to this analysis. The first is the role of Judeo-Christian traditions in European geohistory. The second is the question of the secularization of nature. Both themes are based on the supposition of the gradual loss of the sacred—the sacred history of the earth and of nature with the rise of geohistory and natural history, respectively. I will complicate these suppositions at various levels. First, on the question of Judeo-Christian geohistory, historians have examined the significance of biblical traditions of sacred history within deep history but almost exclusively within medieval and early modern European traditions.[3] It is problematic to construe sacredness within any particular geographical configuration or through a singular religious paradigm. Sacred ideas of the earth were not confined to the Christian worldview and adopted various non-Abrahamic metaphors and motifs. At the same time, biblical themes of sacred geography played important roles beyond Europe. As I just indicated, biblical themes redefined Indian sacred geography. In the eighteenth century, early Orientalist scholars, in their search for the Judeo-Christian roots of Hinduism, studied Hindu mythological texts for references to Mosaic histories and in the process ascribed powerful sacred imageries, derived simultaneously from biblical and Hindu lineages, to the Indian landscape. In the process, these biblical motifs were also imprinted on Hindu iconography, which led to the retention of diluvial themes within Indian geohistorical thinking and the coexistence of science and mythology in India long after they had been discarded in Europe. India similarly appeared to be a sacred land within early Islamic imagination.[4]

The point is not whether these sacred traditions were Western, modern, traditional, vernacular, Hindu, Islamic, or Christian but whether even any particular tradition contained diverse epistemological and ontological positions. For example, as we shall see, Orientalists such as Jones and geologists such as Falconer, in their own distinct ways, rendered Indian mythological texts hermeneutically flexible and mutative. This tradition shared its roots with both Indology and evolutionary theory in the Orientalism of the Asiatic Society of Bengal and the evolutionary studies of Indian fauna through fossils, respectively. These processes had a critical influence on the way late nineteenth-century Indologists engaged with Indian nature and mythologies.

The existence of such hybrid intellectual traditions then raises questions about the secularization of nature, particularly with the rise of natural history. The birth of natural history and the secular pursuits of nature have been dominant themes in the historical study of early modern nature. The rise of European commerce, from the seventeenth century, defined a new era in empiricism, objectivity, and European encounters with various cultures of natural history.[5] Scholars have argued that in this period there was a gradual move away from viewing nature as the work of God and the study of the natural world and the laws of nature in all their diversity as an exploration of divine creation. By the seventeenth century, natural history was pursued widely in Europe, in Spain, Germany, Italy, the Netherlands, England, Scotland, and France, which accompanied a shift from philosophical reasoning to empirical investigations of nature.[6]

Modern geographical knowledge has been seen as a product of the Enlightenment.[7] Scholars have suggested that the gradual eclipse and marginalization of Christian philosophy during the late medieval period and the rise of humanism in the Renaissance created a space for European natural history.[8] According to others, the process involved not just the loss of a Christian motif or the notion of the divine mind but of spirituality itself. According to Michel Foucault, natural history, the cornerstone of the Enlightenment, was founded by the loss of cultural semantics, legends, and fables from the history of nature. The preceding "unitary fabric" of history depicted the antiquity of plants or animals in concomitance with the entire semantic network of stories and legends that were associated with them. The Enlightenment stripped off the associated semantics, myths, and legends, and nature became "a dead and useless limb."[9] It is at this loss of cultural semantics that Bruno Latour locates the modern understanding of nature: a fundamental element in the emergence of the modern idea of nature was the separation of nature and culture. The very idea of nature, engendering a separation of the human and the nonhuman, according to him, is a modernist construct.[10]

Historians have shown that this secularization of nature was a much more indefinite process, as ambiguous as the relationship between science and religion itself. They have also traced the gradual secularization of geohistorical thinking throughout the nineteenth and twentieth centuries.[11] Sacred geography emerged as an important theme in modern scholarship with the realization that religion and spirituality defined the early modern understanding of nature. Simon Schama, in his *Landscape and Memory*, examines the lost cultural relationships with nature in Europe. He invokes the deep Western religious, mythological, and cultural connotations imbued in the three elements of nature: wood, water, and rock. Although he poignantly reminds us of the cultural imagery of nature, his work is also a nostalgic yearning for the same, a search for a lost world of nature in the age of grave materialistic ecological destruction. Portrayals of the complex relationship between science and religion in Europe, therefore, reside in the sense of *loss* of the sacred in nature.

In contrast, Edward Said's article on Jerusalem presents a more political view of modern sacred geographical deliberations beyond Europe. Said shows how the site is layered with sacred and political meanings that are constantly imagined, manipulated, and invented by different communities.[12] Tariq Jazeel has produced a similarly political reading of sacred sites in postcolonial Sri Lanka and highlights the problems in creating hegemonic majoritarian ideas of sacred geography.[13] Jazeel's work is also part of a genre of literature on sacred geography in Asia that uses the term mainly in studies of sacred or pilgrimage sites.[14]

This leads to our third problem in understanding how the sacred resides within the secular. The above studies provide us with significant insights into the endurance and even resurgence of sacred geography, particularly in Asia. However, they do not necessarily explore the complexities of the idea of the "sacred" itself, how in several parts of Asia and Africa, ideas of the secular and the sacred in nature, not just the physical sites themselves, continue to cohabitate. In India, for example, there is a juxtaposition of geographies that are simultaneously sacred and secular. The same River Ganga that is widely revered among Hindus as sacred is also one of the most polluted rivers in India. Reverence for the sacred river is not tempered by its toxic pollution. This contradiction can be explained by the fact that in Hindu practices, there is a fundamental interspersion of the spiritual and the material that allows such incommensurability to coexist. In India, one is constantly faced with this entwining of the sacred and the secular or the ethereal and the physical. The same river is simultaneously a material and natural object as well as a spiritual abode. Ganga is a sacred river to bathe in and feel purified by. It is also a body of water to wash one's clothes in and dump the daily sewage into.[15] Thus, we must consider not only the existence of various religious heritages of nature but also the contiguity of sacred and secular ideas of nature.

Therefore, in this chapter I focus on the retention and coexistence of sacred and secular geohistories in India to illustrate that the deep historical redefinition of myths in India took place at different stages. It started with the introduction of biblical history to India in the late eighteenth century, which was a significant part of the early Orientalist natural and historical imagination. In the process, biblical geography borrowed significantly from Hindu historical and geographical traditions and therefore gained an important afterlife in India, even after its demise in Europe. It also provided Indian nature and natural imaginations with a new sacred theme that was both biblical and Hindu. This then led to the birth of a distinct relationship between Orientalist philology and geology, forged by the use of mythology. Finally, this Orientalist sacred geography acquired a deep naturalistic dimension through geological reinterpretation in the nineteenth century. I explore the latter through two objects that have enjoyed dual lives as myths and fossils: the shaligram and the tortoise.

Sacred Orientalism: The Search for the Asiatic Flood

In 1783, as Jones started his journey to India from Britain to take up his position as a judge on the Supreme Court of Calcutta, he drew up his list of "objects of enquiry during my residence in Asia." It included topics such as "the history of the ancient world" and "traditions concerning the deluge, &c."[16] Jones was inspired by Bryant's work. In 1777, he spent two days with the scholar and commented, "I . . . am wonderfully diverted by his book."[17] Jones's writings in India were the first sustained reading and translation of Hindu Sanskrit texts by a European scholar. He found similarities to classical languages, which to him reflected the similar origins of the people who spoke these languages. This was the basis of his comparative philology. He believed that all great civilizations had common roots but that faith had a clear Judeo-Christian telos. Jones believed that the discovery of the site of human origin in the Middle East was a vindication of Genesis, and thus Mosaic history.[18] His study of Sanskrit and Hindu religion was directed toward vindicating the Mosaic history of the primitive world. As P. J. Marshall has shown, Jones sought to put to rest the doubts in the minds of European theological scholars by showing that Indian antiquity did not begin before 2000 BC or precede Genesis. He even claimed that Hindu myths were versions of Genesis.[19] Therefore, as Urs App has argued, Jones's project was primarily theological, not linguistic or ethnological,[20] which opened up religious conversations between the reading of the Bible and the non-Abrahamic religious texts of Asia. Jones and other Orientalist scholars sought to return Judeo-Christianity to its universal and global antiquity, as religions such as Hinduism claimed to be older and have independent roots. This early Orientalism was distinct from the more political Orientalism discussed by Edward W. Said, in which he argues Europe reconstructed the Orient politically, culturally, and ideologically.[21] This Orientalism was defined more by religious ideology than political hegemony.

In one of his earliest works on Hindu religion, Jones asserted that all ancient religions had common origins in Mosaic history. He was convinced that "a connexion subsisted between the old idolatrous nations of *Egypt*, *India*, *Greece*, and *Italy*, long before they migrated to their several settlements, and consequently before the birth of MOSES; but the proof of this proposition will in no degree affect the truth and sanctity of the *Mosaick* History, which, if confirmation were necessary, it would rather tend to confirm."[22] This meant that his reading of Hindu mythology mixed with what he considered rational history. In his reading of Indian mythology as historical evidence of the references to the flood, the premise of "true history" was fundamental. He started his inquiry by noting that in *all* mythology "historical, or natural, truth has been perverted into fable by ignorance, imagination, flattery, or stupidity."[23] That "true history" to which he sought to conform Hindu mythologies was in itself Judeo-Christian mythology. Jones elaborated that the true history of the world is divisible into Diluvian ones, the times preceding the deluge, and those succeeding it "till the mad introduction of idolatry at Babel."[24] He suggested, "We may infer a general union or affinity between the most distinguished inhabitants of the primitive world, at the time when they deviated, as they did too early deviate, from the rational adoration of the only true GOD."[25] Therefore, to Jones, Christian mythology was the rational, true, and universal history. His project was therefore not just a comparative and global reading of mythology, as has been suggested.[26] His universality had a provincial ambition: to locate Hindu mythologies within Judeo-Christian "history."

In identifying traces of Genesis in Hindu mythology, Jones adopted two principle methods. He depended heavily on the Puranas, which he regarded as the "epitome of first Indian history."[27] He also adopted Vishnu's avatars as the mode of incarnation. Within Hindu mythology, avatars referred to the descent of a deity, chiefly Vishnu, to the earth in incarnate forms. European theologians, travelers, and missionaries had deliberated upon the similarities between Hindu avatars and Christian incarnation since the seventeenth century.[28]

Bryant had briefly noted these similarities between incarnation and Vishnu's avatars deriving from seventieth-century Dutch sources.[29] Jones provided one of the first modern systematic comparative analyses of the two. He appreciated the symbolic and figurative modes of writing in the Puranas. In his discourse "On the Origin and Families of Nations," he suggested that rather than seeing the Puranas as a "mere assemblage of metaphors," the need was to appreciate the "symbolical mode of writing adopted by eastern sages, to embellish and dignify historical truth."[30] This would allow scholars, he believed, to "produce the same account of the *creation* and the *fall*, expressed by symbols very nearly similar, from the *Purânas* themselves, and even from the *Véda*, which appears to stand next in antiquity to the five books of MOSES."[31]

For his depiction of the avatars, Jones drew heavily from the Puranas (such as the Matsya [fish] Purana), for their depiction of the deluge and the ten avatars, which he compared with the deluge described by Moses.[32] He also found resemblances between Zeus, or Jupiter, and the triple divinity of Vishnu, Shiva, and Brahma.[33] In his "Discourse on the Hindus," Jones established a link between Vishnu, avatars, and the deluge.[34] According to him, the first three avatars, or "descents of *Vishnu*" (fish, tortoise, and boar), were the allegorical representations of the "universal deluge."[35] He thus deduced that the second, or "silver," age of the Hindus was subsequent to the dispersion from Babel.[36] The imagery of the boar and tortoise sustaining the globe, which had been affected by violent assaults by demons as the gods churned the sea, suggested to Jones that each of the other avatars had at least one reference to the deluge.

Such hermeneutic and allegorical reading opened up almost unlimited (and in some respects fantastic) possibilities in the Puranas for Jones and his fellow Orientalists. Jones now found an entire Purana dedicated to the description of antediluvian history.[37] This reading also aligned Indian myths with geohistorical themes. He found stories referring to cities inundated by "eruptions from burning mountains, territories laid waste by hurricanes, and whole islands depopulated by earthquakes" and "the first propagation and early dispersion of humans in separate families to separate places." Jones described how "three sons of the just and virtuous man," Noah, then dispersed and populated the world. Children of Japheth became the Tartarians, children of Shem became Arabs, and those of Ham became Indians.[38]

Jones's adoption of the Puranas and avatars as conduits of his comparative theology allowed for both the plurality and singularity of the Judeo-Christian tradition. He suggested that although these traditions could be found in diverse religious traditions, they conformed to the central tenets of Judeo-

Christianity. Jones also placed Indian classical texts adjacent to European theology. Avatarism allowed Jones to situate as Hinduism both concomitant to and part of Christianity. Beyond theology, the Puranas could now be used to extract different meanings that their symbolic writings allowed to serve different ideological purposes. The use of avatars and Vaishnavic symbolism created the meeting grounds of science and mythology in Indian history and of future attempts at establishing the geohistorical truth and validity of the avatars.[39]

Jones's reading of the Puranas was different from contemporary European theological and geological deliberations around Genesis, which sought to establish the validity of the theological text as true history.[40] Jones was not primarily interested in Hinduism, avatars, or even the validity of the Puranas as historical texts, beyond their relevance to contemporary debates on Christianity. Like the savants of Europe whom he followed, his faith was in Genesis, not in the Puranas, and he sought to prove the authenticity of Genesis by using the Puranas as evidence. This obverse historicism, the use of Indian mythologies as evidence in proving *other* truths, remained a feature of the relationship between science and mythology in India, whereby scholars such as Falconer referred to myths to establish geological facts. Jones's comparative study of the Puranas and Genesis did not necessarily link geology with Hinduism, but it produced the coda for both the references to Indian mythology made by later geologists and the links to be made between natural history and Hindu mythology.

Edward Moor started his career in India as part of the East India Company army and then became an Orientalist scholar. He sought to familiarize English readers with the Hindu religion in his *Hindu Pantheon* (1810). He too referred to the Puranas for their evidence of the deluge. He found references to the flood in the stories of the second avatar, the tortoise. According to him, Vishnu assumed the form of the tortoise and then sustained Mount Mandara, placed on his back to serve as the axis on which the gods and demons, using the great *vasuki* snake as a rope, churned the oceans for the recovery of amrita, the drink of immortality.[41] Another Orientalist scholar, Thomas Maurice, published *The History of Hindostan* in 1795. The first volume was dedicated entirely to "Indian accounts" of the "general deluge" and antediluvian and postdiluvian history. His fascinating collection of iconographies of Vishnu's avatars closely followed Jones's narrative strategy and sources in a more enlarged and elaborate discussion of each of the first four avatars, with references to the deluge and the Judeo-Christian origins of Indian antiquity.[42]

Curiously, the images of Vishnu's avatars in Maurice's *History of Hindostan* have clear European landscape backgrounds. The "Varah avatar," for example, features European castles within the earth sustained by its tusks (see, for example, fig. 3.1).[43] This is particularly striking because Maurice complained about the lack of perspective in such Indian paintings and maintained that these images had been reproduced without any alteration: "The three last engravings [first three avatars, fish, boar and tortoise] are *fac similes* of the Avatars, as they are painted in the pagodas of a people who are utter strangers to perspective; and it was therefore thought improper to alter them. Perhaps the very eccentricity of the design, as it undoubtedly stamps their originality, may procure them admirers."[44]

Despite Maurice's claim, these images of Vishnu's avatars had complex historical trajectories. Thomas Trautmann has referred to the massive statue of Jones at St Paul's Cathedral in London (by John Bacon) in which Jones is wearing a toga and has Manu's "Institutes" in his hands. At the front of the pedestal of the statue, there are two figures on either side holding a circular plate with an engraving of the avatars of Vishnu.[45] Trautmann characterizes this as the last instance of Hindu iconography being used in the "scientifically elicited proof of the truth of Christianity." He contends that later in the nineteenth century, as Indophobia set in, such eclectic religious spaces faded away. He describes the installation of the statue as "Christian Indomania's brief moment."[46] Apart from being an image of a Hindu god in a church, the engraving is that of the tortoise (*courma*) avatar, which is startlingly similar to the *courma* avatar in Maurice's *History of Hindostan*.

Jones's statue was created in 1799, just a few years after the publication of Maurice's book, and the image could have been adopted from it. Alternatively, it could refer to the wider circulation of these images in Europe. The intriguing European landscape backdrops to the avataric images, in fact, indicate the long historical lineage of linking Vishnu's avatars with the deluge and biblical imagery. The images used in Maurice's book circulated in Europe and appear in the early eighteenth-century work of the French engraver Bernard Picart (in his *Cérémonies et coutumes religieuses de tous les peuples du monde*).[47] Interestingly, these images had appeared in Europe much earlier, for example, in Philip Baldaeus's work, which was published in 1671 (fig. 3.1).

Baldaeus was a Dutch minister who traveled to Sri Lanka and India in the seventeenth century and commented on the similarities between avatars and the biblical deluge.[48] European missionaries referred to Vishnu's avatars since the seventeenth century, particularly the Dutch missionary Abraham Rogerius in his book *Open Door to the Hidden Heathen Religion* (published in 1651).[49]

Fig. 3.1. The boar avatar. Philippus Baldaeus, *A True and Exact Description* (1745), © British Library Board 455.f.3, following 748. The same image appeared in Maurice's *History of Hindostan*, 1:560.

The exact origin of these images remains obscure. Daniel Bassuk suggests that almost all of Baldaeus's work was plagiarized from an anonymous manuscript written in the mid-seventeenth century.[50] He does not specify which. It could be the "Deex-Autaers" (Ten avatars), a manuscript the Dutch painter and publisher Philip Angel had acquired in Surat around 1650 on his way to Batavia from Persia.[51] It is probable that Picart adopted these images from works of Baldaeus or other Dutch artists. It is useful to remember that Picart moved to Amsterdam in 1711.[52]

While the exact origin of these images or the insertion of European landscapes remains unclear, three points emerge here. First, these images of Hindu iconography were not of singular Indian or Hindu origin, as claimed by Maurice, but contained layers of myriad cultural and religious motifs imprinted on them in different ages and historical settings. Late eighteenth-century Orientalist scholars saw these highly textured images as evidence of authentic Hindu representations of the deluge. Second, avataric images circulated across Europe and Asia, and across different religious configurations, at least from the seventeenth century. Ideas of the deluge along with others such as the Mount Meru or the Aryavarta, which I will discuss later, traveled across geographical and religious borders in the early modern period.

Finally, the epistemological discontinuities in the global circulation of these images from the early modern period are critical to understanding their precise historical significance. These images acquired new meanings within late eighteenth- and early nineteenth-century sacred Orientalism when they were interpreted by Jones and reprinted by Maurice. Baldaeus did not suggest in the seventeenth century that Hinduism was part of the biblical corpus, as Jones later did. Baldaeus merely observed that Hindus might have heard stories of the deluge that circulated widely in the medieval and early modern Eurasian world.[53] It was only with the late eighteenth-century preoccupation with universal and "true history" that these images appeared as part of the Abrahamic religious corpus. As recent research has shown, the late eighteenth-century British reading of Indian theology had much deeper engagements with the philosophy and origin of Hinduism, compared to the earlier readings.[54] In the process, paradoxically, these images also appeared to be more authentically Hindu and lost their earlier hybrid lineage. It was through this alignment between European theological deliberations and Hindu avatarism, as we shall see, that these Puranic images and narratives were also linked to earth histories.

The Sacred-Natural World of Orientalism

To understand how the Puranas and avatarism were linked to geohistory, it is important to appreciate the early Orientalist identification of Indian geography as a simultaneously natural and mythological entity. Because of its biblical motif, Jones's reading of Indian mythology had strong geographical overtones. At the same time, his description of Indian geography was also acutely mythological. To give an example, his interest in the Himalayas started with his desire to learn about Buddhism and Tibet, which some of his contemporaries believed was the cradle of human civilization.[55] He had learned about

Tibet from the work of two Italian Capuchin missionaries, Francesco Orazio della Penna, who produced an 800-page volume of the mythology and history of Tibet, and Agostino Antonio Giorgi, who wrote *Alphabetum Tibetanum* (1762).[56] These Catholic monks settled in Lhasa during the early to mid-eighteenth century. Their books were intended to assist future missionaries visiting Tibet. They had biblical motifs similar to Jones's, depicting Tibetan religion, rituals, history, and geography as derived from the Manichaean heresy, the belief that the spirit could be released through ascetic practices. Their description of the Himalayas carried those ascetic sensibilities.

Jones wrote about the Himalayas in his translation of the Sanskrit hymns dedicated to the Ganga (referred to as both a river and a goddess). Before narrating the hymns, he added "geographical notes" on the rise of the river in the lofty mountains and its subsequent journey through the northern plains. He believed that without these notes, the hymns would appear "obscure."[57] Despite his intention of providing geographical clarity, these notes seamlessly assimilated the sacred and the natural. Jones referred to Sanskrit and Chinese sources, as well as Ptolemy and Giorgi, while painting a lyrical and mystical geographical world in which he described the birth of the Ganga from the forehead of Lord Shiva, whom he described as the Jupiter Tonans and genitor of Latin. The river's point of entry into Hindustan was at Sambal, which according to him was the Sambalca of Ptolemy. In the northern Indian plains, Ganga was "joined by the *Calinadi*, and pursues her course to *Prayaga*, whence the people of *Bahar* were named Prasii, and where the *Yamuna*, having received the Sereswati below *Indraprest'ha* or Delhi, and watered the poetical ground of Mathura and *Agara* [Agra], mingles her noble stream with the Ganga close to the modern fort of *Ilahabad* [Allahabad]."[58] According to him, the Himalayas, which he described as the "powerful monarch," derived their name from the Sanskrit word *haimus*, which means "snowy" and was the same as the Greek *hamus*.[59] He referred to Giorgi to describe the Gandak River in Tibet, which according to him the Greeks knew by the same name and was believed to be filled with crocodiles of "enormous magnitude."[60]

The point here is not whether these geographies are real or mythical. The problem is in understanding how the sacred/mythical and the natural coexisted. Even "real" places such as Jerusalem are constantly imagined, invented, and defined at a remove from their physical reality.[61] Ian Mabbett has shown that the idea of the cosmic Mount Meru, as it entered Indian mythology from Babylonian during the Seleucid Greek contact with the Mauryan Empire around 300 BC, acquired its "third dimension" in conversations with Hinduism and Buddhism. This new dimension could not be mapped; "it pierced the heavens; in piercing the heavens, it transcended time as well as space; in transcending time it became (in Mus's sense) a magical tool for the rupture of plane. . . . Meru is not, we must recognize, a place, 'out there,' so to speak. It is 'in here.' "[62] Jones drew from this sacred-natural tradition in his discussions of Meru. While describing the Puranas as antediluvian texts, he proposed that Hindus had come to India from an original homeland, a land of mountains in the north, at the source of the River Ganga at the summit of Meru. There Meru was both a mythical and geographical place. Jones provided the geographical context to suggest that the remainder of the Indian subcontinent was at that time insulated by the mountains.[63] In this eighteenth-century mapping of sacred geography, Meru, the Ganga, and the Himalayas appeared as both real and imagined entities. The coexistence of the sacred and the secular is critical in understanding the deep historical interpretation of Indian mythology and the mythological interpretation of Himalayan fossils in the nineteenth century.

Francis Wilford was the first Orientalist scholar to provide geographical mappings of Indian mythology. A lieutenant in the Bengal Engineers until 1794, Wilford was an assistant to the surveyor general of Bengal between 1786 and 1790 and, from 1794, secretary of the Sanskrit College in Benares. In the 1792 issue of the *Asiatick Researches*, he explained how his project to write an "entirely *geographical*" account of India and Egypt became infused with mythology, where the lines between the real and the imaginary were blurred: "I was obliged, therefore, to study such parts of their ancient books, as contained geographical information, and to follow the track, real or imaginary, of their Deities and Heroes; comparing all their legends with such accounts of holy places in the regions of the West, as have been preserved by the *Greek* Mythologists, and endeavouring to prove the identity of places by the similarity of *names* and of *remarkable circumstances*."[64] Like Jones, he too believed that Indian myths were "inconsistent and contradictory" and that Indian "history" was "involved in allegories and enigmas, which could seem extravagant and ridiculous." Therefore, he, with the help of Brahmin pundits, sought to give new meaning, rationality, and most importantly, a natural basis to Hindu mythology.

It is true, as suggested by Nigel Leask, that Wilford "construct[ed]" and metamorphosed Hindu geography from cosmographic representations.[65] He attempted a rather simplistic and superficial imposition of sacred maps on geographical sites. His "Hindu geography" was based more on conjecture regarding similar names and places across cultures than on any deep textual research or geographical exploration.[66] He linked Sanskrit terms phonetically with Latin and Greek ones: Mount Caucasus was for him Hindu Kush; Mount Ararat was "Aryavarta" or "Aryawart."[67] Wilford's "conjectural etymology" was also based on fake stories. He discovered, to his dismay, that his Hindu interpreter had taken creative license. In his introduction to his essay "Sacred Isles of the West" in the 1808 edition of the *Asiatick Researches*, he admitted that his assistant had substituted words and inserted pages into manuscripts and had even made up legends.[68] Wilford remains best known for adopting means that have been referred to as "pious fraud."[69]

Despite this dubiousness, Wilford, along with Jones and Maurice, also opened up the possibility of imprinting imagined geographies over real ones in South Asia. While his accounts and sources were discredited, his methodology—mythology-focused geographical and historical inquiry—remained valid. Most importantly, it presented the blueprint for later scholars to write their Hindu versions of sacred geography along very similar lines. In the early twentieth century, Indian scholars identified Meru and Aryavarta that Wilford had referred to, from the same scriptures as geographical evidence of their Vedic heritage.[70] While increasingly marginalized in Europe, sacred geography found a new lease of life in late nineteenth-century India through these various interpretive processes. Later Hindu accounts of Aryavarta, Rama Setu, or the mythical river Saraswati borrowed their methodology from these traditions. However, they tended to draw, as we have seen in the case of Max Müller and the "Sarasvati," more from the Vedic sources than from the Puranas, which were increasingly regarded as too allegorical. The growing "Indo-Aryanism" of the late nineteenth century among European and Hindu scholars such as Müller, Dayananda Sarasvati, and B. G. Tilak derived from nineteenth-century European Aryan race theory and referred to Vedic sources to claim the Aryan origins of Brahmins.[71] Kedarnath Datta was among the Hindu scholars who aligned Hindu avataric ideas with the racial evolution of the Hindus, suggesting that the British and the Bengalis were brothers, as they belonged to the same Aryan race.[72]

These reinterpretations of Indian geographical traditions were accompanied by another significant shift in the understanding of sacred geography in the nineteenth century: the growing secular trend within Orientalism. Javed Majeed has traced the transformations in British imaginations of India from Jones in the late eighteenth century to the political philosopher James Mill in the nineteenth century. He sees Mill's rejection of the earlier Orientalist, mythology-based, "ungoverned" imaginations of India as necessitated by the politics of governance that preoccupied him.[73]

Jones and Mill were not necessarily the only opposing ends of the British imagination of India. It was not just Mill who wanted imaginations about India to be "governed." Orientalism incorporated various opposing purposes and trajectories within itself. By the early nineteenth century, it had shed most of its earlier sacred geographical motives and adopted topographical and commercial ones. For example, the Orientalist Henry Thomas Colebrooke, often regarded as Jones's intellectual heir, took a very secular and political approach to Indian antiquity and geography. His descriptions of the Himalayas had none of the mystic quality that marked Jones's accounts. The Himalayan expeditions during Colebrooke's time were part of British territorial expansion in the region. The triumphs in the war with the Gurkhas in 1814 secured the Garhwal and Kumaon regions of the Himalayas for the company. Colebrooke's accounts of the Himalayas, therefore, had clear commercial and political motives. The explorations of the Himalayas in the early nineteenth century took place within this paradigm of secular geography. A distinct genre of Himalayan accounts developed in the early nineteenth century, when British travelers, rather than depending on ancient texts depicting the lofty mountains, actually explored the Himalayan landscape. Samuel Turner's account of Tibet, published in 1800, for example, was driven by Hastings's ambitions to access the commercial routes across the Himalayas leading to Central Asia.[74] These expeditions, which were distinct from those of Giorgi, Penna, Wilford, and Jones, rewrote the geography of the Himalayas.

Colebrook's work reflected this new commercial and topographical urgency. In his introductory remarks to Moorcroft's notes on his journey to Manasarovar Lake, he pointed out that the journey was undertaken with the motive of "publick zeal" "to open to Great Britain means of obtaining the materials of the finest woollen fabric."[75] Moorcroft himself wanted to explore the horse trade of Central Asia to improve the stud quality of British horses in India in addition to the wool trade. Colebrooke was keen to correct the misconception that the Manasarovar was the source of the River Ganga: "It gives origin neither to the Ganges nor to any other of the rivers reputed to flow from it."[76] He deliberated about the correct heights of the Himalayas that the explorers had now to scale.[77] This is quite different from the way Jones had earlier written about the Himalayas and the Ganga.

Several factors led to the emergence of this secular geographical tradition within Indian Orientalism in the nineteenth century. They developed partly because, as suggested by Arnold, notions of improvement and economy dominated European geographical explorations and imagination in the nineteenth century.[78] The territorial expansion of the EIC in northern India in the early decades of the nineteenth century was one. Arnold has also referred to British attempts at disassociating the scenic nature of India from its vernacular connotations (for example, making distinctions between the beautiful and serene Ganga and the "ghastly" Hindu practices on its banks).[79] He argues that in the early nineteenth century, despite the EIC's efforts to disassociate itself from it, there was a dispersal of evangelical ideas and aesthetics among British residents in India. The evangelical attack on Hinduism from the early nineteenth century severed any possibility of appreciating Hindu imaginations of Oriental nature.[80] The same Hindu idols, gods, and iconography that had opened up different possibilities earlier now appeared "obscene": "Nature in India thus seemed complicit in the 'horrors' of Hinduism."[81]

Missionaries working in India in the nineteenth century, who were keen to demonstrate the incompatibility of Hinduism with Christianity, rejected Jones's universal Mosaic history. William Ward, the English Baptist missionary based in Serampore in Bengal, included a summary of Vishnu's ten incarnations in his ethnographic work on Indian history, literature, and mythology published in 1815. He described these traditions as "contemptible," without any deep historical truth. Ward wrote that "Mŭtsyŭ" (the fish avatar) appeared following a "periodical destruction" of the universe when the Vedas "remained in the waters" and needed retrieving. As "Kŭchyŭpŭ" (the tortoise avatar), Vishnu "assumed the form of a tortoise, and took the newly created earth upon his back, to render it stable," adding, "the Hindoos believe to this hour the earth is supported on the back of a tortoise." "Vŭrahŭ" (the boar avatar) appeared after this destruction of the world, "when the earth sunk into the waters" and needed to be saved. Ward ridiculed the idea of the "earth, with all its mountains, &c. &c. made fast on the back of a turtle, or drawn up from the deep by the tusks of a hog!"[82]

Orientalism in the nineteenth century had moved away from its theological premises of writing a universal Mosaic history. By then, the Bible as the original source of religious civilization was reduced to a moral and symbolic source within Orientalist scholarship.[83] Along with that, the rise of Indology as a distinct discipline in the nineteenth century meant that the study of Indian classical texts gathered a secular meaning in itself and lost the need to refer to biblical themes. The earlier traditions of sacred geography disappeared from Indology at the same time that Indology adopted Hinduism as its primary intellectual preoccupation. The secular naturalism that operated within the emerging Indological tradition, as evident in Colebrooke's work, was also paradoxically invested in the study of the Hindu religion. Thus, although the secular geographical tradition of the nineteenth century carried religious and cultural messages, these messages were now more clearly identified as "Hindu" traditions, rather than as a hybrid of Hindu, Buddhist, and Abrahamic religions, as was characteristic of the earlier period. In the process, while Hinduism retained the earlier sacred geographical themes, it also accommodated the new secular geographical implications. Hinduism appeared as a singular religious tradition that contained the deep cultural and geological legacies of India.

The coexistence of these secular geographical and mythological imaginations led to the resurgence of deluge imagery within Indian nationalist literature in the early twentieth century, mixing ancient texts with references to geological landscapes.[84] The sacred geographies of Aryavarta, Mount Meru, and Rama Setu acquired clear geographical and even geological motifs by the end of the nineteenth century. Wilford had first proposed Aryavarta as a geographical region in Central Asia.[85] In 1912, the Brahmin Sanskrit scholar from Pune, Narayan Bhavanrao Pavgee, published *The Vedic Fathers of Geology*.[86] The book provided a geological account of the Aryan origins of Hindus. Pavgee argued that the Vedic "forefathers" were early geologists. He traced symbolic representations of Azoic, Paleozoic, Mesozoic, Tertiary, and Quaternary epochs in Vedic texts to narrate a geological story of the homeland of Vedic ancestors in the Arctic region, in the land of Aryavarta.[87] From these geological insights, he also sought to establish the antiquity of the Vedas in the Tertiary period, 80,000 to 240,000 years ago.[88] In doing so, Pavgee borrowed heavily from various metropolitan works on Indology and evolution by Max Müller, Monier Monier-Williams, and Louis Jacolliot.[89]

Other scholars had similarly incorporated geological information into the Indological mapping of Indian antiquity. Two years after the publication of Pavgee's book, in 1914, the Ayurvedic physician and publisher Binodbihari Roy of Rajshahi in eastern Bengal published a detailed account of the geological features of "Meru-desh" (the land of Meru), which he believed was situated in the Arctic and was the original homeland of the Aryans. He described its subsequent destruction during the Ice Age and the gradual southern migration of Aryans to "Su-Meru," which was closer to India.[90] Meru, or Mount Meru, has a complicated geomythical lineage. Although identified by Wilford as a biblical reference in Hindu texts, it is believed to have entered Hindu texts and

Indian mythology through the Seleucid Greeks during the Mauryan period, between 320 and 180 BC.[91] While Pavgee and Roy drew from Puranic and Vedic sources, following the conjectural methodology of Jones and Wilford, they did not share the latter's sacred geographical imagination.

Vishnu's Fossils

The gradual insertion of geological imagination into Indian sacred geography is evident in the European discovery of Himalayan ammonite fossils, or shaligrams (also known as salagrams, silagrams, or saligrams). Shaligrams are unique in their cultural and geological heritage: these ammonite fossils of extinct marine animals were venerated by Hindus as representations of Vishnu's avatars. Shaligrams have thus been used, as I will examine, as examples of Indian geomyths, as evidence of Indian evolutionary thinking. However, it is not clear how the fossil folklore and the avataric mythologies around shaligrams were formed. Within nineteenth-century Orientalist scholarship in India, shaligrams were often placed in two domains: as artifacts of Hindu avatarism and ammonite fossils. Both Orientalism and evolutionary science evolved through the nineteenth century, along with ideas of the geological formation of the Himalayas, where these fossils were found. Consequently, the various identities of shaligrams as objects of Vishnu's incarnation, symbols of Hindu evolutionary thinking, or evidence of the Himalayan formation from beds of the submerged Tethys Sea were fused. Shaligrams therefore offer an important insight into how Vaishnavism and geology came together in the nineteenth century. They also provide us with the opportunity to examine the theme of geomythology.

From the eighteenth century, Europeans collected shaligrams mostly as cultural artifacts and referred to their connections with Vishnu and his avatars within Hindu cosmology. The EIC soldier and administrator William Henry Sleeman served in the Anglo-Gurkha War between 1814 and 1816. During this expedition, he observed the reverence among Hindus for the local shaligrams, which he believed had been washed down the rivers from the central Himalayas. He viewed this veneration with a degree of sarcasm. Sleeman narrated an episode when during the Nepal war, a certain "Captain B" brought back to the camp four or five shaligrams he had taken from the hut of a priest belonging to the Nepalese forces. The captain then asked for a large stone and hammer and "calmly" cracked the stones like "walnuts" to examine their contents, while the local Hindus watched in a state of deep anxiety, apparently expecting "the earth open and swallow up the whole camp."[92]

Others like Maurice were more sympathetic to Hindu sentiment regarding these stones. He noted that Hindus venerated them for their "mystical virtue." He believed that they were found in the river "Casi" (Koshi), a branch of the Ganga in the Himalayas. According to him, Hindus believed that shaligrams appeared in nine shades, each of which represented an incarnation of Vishnu.[93] The Brahmins believed that a small worm formed the spiral cavities inside the stones, which while working its way inside the stone, prepared the abode of Vishnu in its bosom. Some saw in these spiral lines the image of Vishnu's *chakra* (a discus or mystic circle, held by Vishnu, suggesting the passage of time).[94] According to Hindu legend, the black ones represented the avataric transformations of Vishnu into Lord Krishna (the dark-colored eighth avatar of Vishnu), and those tinged with violet represented Vishnu's "angry incarnations," the "man-lion," or the "Rama avatars."[95] He noted that worshippers carefully preserved the stones on the altars of Vaishnav temples. They were covered in white linen, washed every morning, anointed with oils, and perfumed.[96]

Wilford, in his long essay on the ancient geography of India published in the *Asiatick Researches,* wrote about the shaligrams found in the Gandak River, a tributary of the Ganga, in the remote central Himalayas. He came up with various names for the shaligram, such as the Sailgaram, Saila-chacra, and Gandaci-shila, from its religious associations. According to him, the name "shaligram" came from a village called Sailagram, situated in the snowy mountains. The origin of the mountain on which the village was situated was itself associated with one of Vishnu's legends, according to which Vishnu hid in the stone in the shape of a worm in this region to escape the power and influence of Saturn. Once the scourge ended, he ordered this stone, his sanctuary, be worshipped.[97]

Such references to shaligrams as artifacts of Vishnu's abode on earth became commonplace in early nineteenth-century Orientalist texts, often alluding to their mystical qualities. Edward Moor, in his *Oriental Fragments,* noted that Hindus had "a mystic reverence for lithic forms" such as shaligrams. He suggested that each type of shaligram represented an avatar.[98] He added, "Volumes have been written on its mysteriousness and virtues." Its use had similarities to Christian sacraments and ceremonies: "Several ceremonies are uncompletable [*sic*] without one. In death, it is as essential an ingredient in the *viaticum,* to at least one sect of *Vaishnava*—perhaps to many sects—as is the *oleo santo* of Papists. The departing Hindu holds it in his hands—an easier, and less disturbing, and less unpleasant process than the greasings of the dying Papist."[99] Shaligrams were rarely regarded as fossils in these references.

Fossils were not always paleontological objects. In medieval Europe, fossils, including ammonites, were often regarded as religious objects.[100] Snakestones, which are also ammonite fossils, appeared as part of the myths about snakes in Europe.[101] The fossilized bones of hyenas that William Buckland found in the Kirkdale Cave were assumed by the local quarrymen who first found them to be relatively fresh remains of cattle that had died in a recent epidemic and used them to fill potholes.[102] Similarly, Hindus used them as objects of worship, traded them as items of commerce, or used them as charms and medicines.

In their original sense, fossils were objects "dug up" from the earth. Thus, they had inherent geological connotations even when their prehistoric qualities were not evident. By the late eighteenth century, European naturalists identified them as organic remains and deliberated upon their distinction from living creatures. Rudwick shows that European naturalists at this time sought to explain this difference through the three causalities of extinction, migration, and transmutation.[103] However, their prehistoric identities were yet to be established. By the 1820s, the paradigm of "characteristic fossils" associated with William Smith acquired its own prominence, with fossils serving to classify geological strata.[104] William Smith identified ammonites and mussels as fossils in his identification of strata.[105] Previously, the physical characteristics of strata had a more critical bearing on classification, for example, the Cambrian-Silurian (or Sedgwick-Murchison) dispute explored by James A. Secord and the Devonian controversy considered by Rudwick.[106] However, index or zone fossils came to be a critical factor in biostratigraphic correlation. With the development of paleontology, ammonoids were considered one of the most important index fossils because of their distinguishable features, rapid evolution, high occurrence, presence in diverse environments, and wide geographical distribution. Himalayan shaligrams, particularly those in the Gandak River, are now regarded as the most significant evidence of the death of the Tethys Sea, as the Indian plate collided with the Asiatic landmass.[107]

It is important to note that in India in the eighteenth century, British Orientalists and geologists rarely referred to shaligrams as "fossils" even in their original sense as objects belonging to the earth, even when naturalists in Europe regarded ammonites as fossils. This is partly because Europeans at this time did not always dig them up from the earth in the subcontinent. They often collected them as Eastern exotic objects from markets and pilgrimage sites.

The critical transformation in the understanding of shaligrams as geohistorical objects took place in the nineteenth century, when they and their associated myths were linked to the geological exploration of the Himalayas. The large-scale collection of shaligrams as fossils from the remote Himalayas started as the British explored the region for minerals as well as passages to the Central Asian trade routes. Shaligrams also shaped ideas of the geological features of these mighty mountains. Alexander Gerard's expeditions in the Himalayas in 1818 and 1821, following the Anglo-Gurkha Wars of 1814–16, became important for his descriptions of the geography of the mountain ranges and passes and for the ammonite fossils he found there.[108] He brought them to Saharanpur, and they subsequently reached the Asiatic Society's museum in Calcutta. In 1825, Colebrooke presented a note on Gerard's Himalayan travels to the Royal Asiatic Society in London and referred in particular to the ammonite fossils he found.[109] In his statistical sketch of Kumaon region of the Himalayas in the 1830s, George William Traill described the "organic remains" found at the highest altitudes of the Niti Pass in the Himalayas. He referred to shaligrams as "*Chakar Patar*" (literally "wheel-shaped stone") as they were known locally for their resemblance to a wheel.[110] Alongside, he found fossilized bones, locally regarded as *bijlí hár*, or "lightning bones," for their miraculous curative powers.

Based on these discoveries in 1831, J. D. Herbert, assistant to the surveyor general of India, wrote the first detailed geological account of the Himalayas. He followed Saussure's Alpine methods in studying the strata, rocks, and fossils of the Himalayas and referred to ideas of superposition that William Smith had extensively used. He identified shaligrams and "lightening bones" as part of the Himalayan organic geology. His main emphasis was that the organic remains of the Himalayas were distinct from those of the European mountains, which indicated a distinct fauna and flora of the region, shaped by its climate and altitude. Gerard referred to the ammonites to stress the difference in the geological formation of the Himalayas and the Alps (fig. 3.2).[111]

With the gradual recognition of their significance as Himalayan fossils, shaligrams became central to the paleontological collection at the Geological Society of the Asiatic Society of Bengal. The society's museum in Calcutta had collected shaligrams as cultural artifacts (sometimes referred to as *dwáracá-chacra* in the catalog) from the late eighteenth century.[112] Since the 1820s, they came to be more clearly regarded as "fossils."[113] At the same time, shaligrams were sent to London, both as fossils and as cultural artifacts, to two different institutions, the Royal Asiatic Society and the Geological Society.[114] Nathaniel Wallich, while he was based at the Calcutta Botanical Garden, had a large collection of shaligrams that he took with him to England along with his botanical

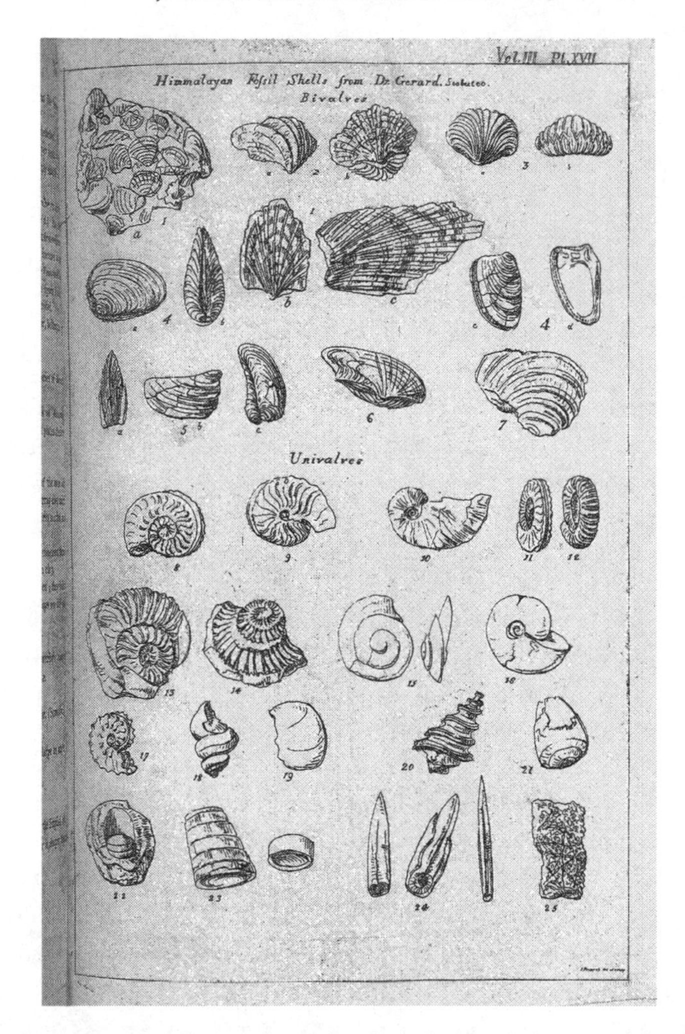

Fig. 3.2. Alexander Gerard's Himalayan fossils, including shaligrams, referred to by Herbert. *Gleanings in Science* 3 (1831), © British Library Board 455.f.3, next to 272.

specimens. They were ultimately deposited at the Sedgwick Museum in Cambridge.[115] The German paleontologist Carl Albert Oppel referred to a specimen of ammonites from the river "Gundock" that were kept in the collection of the Jardin des Plantes in Paris since 1825.[116] As shaligrams, or ammonites, were distributed to different museums in Europe, they generated debates among European geologists about their geographical origin and geological significance.[117]

Yet they remained ambiguous objects. Although they became fossils in the modern sense, as organic remains of Himalayan evolution, shaligrams retained their cultural symbolism. Colebrooke was one of the first scholars to recognize shaligrams as both cultural and paleontological objects.[118] However, the hybrid metaphors he used to present the shaligrams as fossils fused the two identities rather than disconnected them. He referred to shaligrams as "antediluvian molluscs" that had been regarded by Hindus as representations of Vishnu's avatarism. He wrote in his note on the heights of the Himalayas:

> Among the loftiest in that chain is one distinguished by the name of *Dhawalagiri* or the white mountain, situated, as is understood, near the source of the *Gandhac* river called in its early course *Sálagrámî* from the shistous [*sic*] stones containing remains or traces of ammonites found there in the bed of the river and thence carried to all parts of India, where they are worshipped under the name of *Sálagráma*; the spiral retreats of antediluvian molluscs being taken by the superstitious *Hindu* for visible traces of VISHNU.[119]

Colebrooke often referred to shaligrams with this dual identity as both fossils and symbols of avatarism. In an article on the religious ceremonies of the Brahmins, he described the use of the shaligrams in the last rites of Hindus: "A *Salagrama* stone ought to be placed near the dying man, accompanied with holy passages from the vedas or sacred poems repeated aloud in his ears and leaves of holy basil (tulsi) scattered on his head." In the footnote, he added that these were "black stones" found in a part of the Gandak River in Nepal. Here he provided a similar legendary account of its perforations having been created by worms or it being an allegorical form of Vishnu in his various avatars, as the number of spiral curves of the stones represented Vishnu in his various forms.[120] Colebrooke himself, as the president of the Asiatic Society in Calcutta, received several shaligrams, which he donated to the Geological Society of London, with the dual provenance of these being "argillaceous nodules containing Ammonites and called Salagrams, worshipped by the Hindoos."[121] Among the articles donated by the ethnologist and naturalist Brian Houghton Hodgson to the Royal Asiatic Society were references to shaligrams as "fossil objects of worship."[122] Such citations continued in the nineteenth century. The entry "Salagrama" in the *Cyclopædia of India and of Eastern and Southern Asia* (1857) referred to them as both fossils and representations of Vishnu's avatars, with a detailed description of how according to the Hindu religion, their different shapes and colors denoted each avatar.[123] In these persistent references to the dual identity of shaligrams as fossils and objects of Indian avatarism lies the origin of modern geomythologies around shaligrams as representations of ancient Hindu ideas of deep time.

Why did shaligrams retain their sacred associations in the age of prehistory? Scholars such as Colebrook and Hodgson were aware of the distinctions between them as fossils and religious objects. However, the constant juxtaposition of these two identities demonstrates, on the one hand, the Orientalist legacies of Indian fossil studies. On the other, it indicates the merger of Orientalism with geohistory in the second half of the nineteenth century. The latter was aided by the rise of what Mackenzie Brown has referred to as "avataric evolutionism." This conflation was critical to the making of Indian geomyths. The religious identity of shaligrams, in fact, became more significant in the second half of the nineteenth century as they became essential to the Indological discovery of Hindu evolutionary thinking. A renewed interest among Orientalist scholars in Vaishnavism from the 1870s provided a new reading of Indian evolutionary symbols and artifacts. At this point, Vaishnavic avataric imageries were constantly linked with Darwinian evolutionary ideas.[124]

The resurgence of Vaishnavism as a modern religious tradition particularly in Bengal from the mid-nineteenth century—a reenactment of the devotional tradition of the sixteenth century in response to colonial education, scientific ideas, urban growth, and secularism—combined ideas of evolutionary thinking with Vaishnavic philosophy. From the 1890s, Hindu scholars actively merged Darwinian evolutionary ideas with those of Hindu reincarnation. This reinterpretation of Vaishnavic avatarism was commonly accepted in the late nineteenth and early twentieth centuries.[125] Along with this, an extreme form of Indology in the late nineteenth century unfolded chiefly in Europe, in the works of Oxford-based Sanskrit scholars such as Monier-Williams; German Indologists such as Max Müller, Albrecht Weber, and Otto Schrader; and the French colonial intellectual Jacolliot. They identified India as the cradle of civilization and marked a close entente between modern science and Hinduism.[126] These scholars collectively asserted that the major scientific theories of the nineteenth century such as Darwinian evolution had been preempted by Vedic and Puranic scholars. Their faith in the compatibility of Hinduism and modern science was based on the centrality of reincarnation. In these discussions, the Puranas and the avatars became main evidence of embryonic ideas of Hindu evolutionary thinking.[127] By this period, discussions of sacred geography in India had moved away from their original Orientalist and biblical inspiration and adopted evolutionary and geological themes.

Within this new paradigm of Hindu evolutionism, scholars portrayed shaligrams, along with various other fossils, as evidences of Indian geomyths and of Hindu acquaintance with evolution, implying that ancient Hindus were *aware* of their value as fossils. In the process, their identity as cultural artifacts came to be indistinguishable from their prehistorical status as fossils. Recent scholarship continues to portray shaligrams as part of the ancient fossil traditions of India. It links them with ideas of Vaishnavic evolution, indicating thereby that ancient Indians were familiar with the concept of fossils.[128] Others have used the paleontological studies of shaligrams as fossils as evidence of the authenticity of Puranic evolutionism.[129] Scholars have also stressed the physical similarities between the ammonite fossil, or shaligram, and Vishnu's *chakra* to suggest that shaligrams were seen as fossils in ancient India.[130] These ideas have been reinforced by the evidence of similar folklore around ammonite fossils in other parts of the world, including Britain and Germany as well as ancient Greece and Rome.

There is a wider issue at stake here. Scholars have explained the veneration of various cultural artifacts that came to be known as fossils in the early nineteenth century as evidence of indigenous awareness of their deep historical significance. Geomythology has become a key concept for scholars who insist that ancient civilizations were familiar with the observation and collection of fossils. Dorothy Vitaliano conceived of the term to identify how several myths and legends had preconceived geological events.[131] Adrienne Mayor suggested that several myths about objects that came to be known as fossils were evidence of geomythology.[132] She argued that those who formed those myths were the "first fossil hunters" and that the practice of geology, as we know it, started in much earlier historical periods. Such arguments have been applied to India. Mayor herself suggests that folklore regarding dragons in northern India, traceable to Apollonius and Philostratus, is comparable to Scythian legends of griffins as well as Chinese legends of dragons.[133] Others such as Alexandra van der Geer suggest that the Greek *drakones* were analogous to *nagas* in Indian tradition (as well as the *seraphim* in the Hebrew tradition) and that the concept of such half-snake, half-human creatures might have been based on Himalayan fossil ammonites.[134] More radically, these scholars claim that the great epics contained geological truths later discovered by modern science.[135] Some have conducted "ethnobiological research" to trace "ancient scientific thoughts conveyed within the Sanskrit hymns of legendary epics."[136]

This is an imagined intimacy between cultural artifacts and deep history. The interpretation of Vaishnavic ideas of shaligrams and other objects as "fossil mythologies" is problematic. It assumes that Eastern folklore carried allegorical interpretations of fossils without examining how and why Europeans in Asia, since the eighteenth century, employed mythology in geological speculation about deep time. It is necessary to separate various entangled intellectual threads here. To start with, fossils became geohistorical objects in the eighteenth century and prehistoric artifacts in the early nineteenth century through the new interpretive paradigm of deep history. It is not possible to establish that Hindus understood them as fossils without inserting the concept of fossils and prehistory into Indian antiquity. To put it simply, it can be argued that ancient Hindus did not treat shaligrams as fossils because the idea of fossils as artifacts of the deep past is a modern one. Similarly, the links between Vishnu's avatars and Darwinian evolution were created in the late nineteenth century. This is not to suggest that these cultures did not have ideas about the earth, nature, or the emergence and evolution of living forms. However, those ideas were distinct from modern deep history and evolutionary thinking.

Geomyths are appealing because they insert deep historical awareness into non-Western or seemingly non-modern traditions. They imbue cultural ideas with geological facts and in the process provide the former with greater credibility. The treatment of myths as nascent scientific ideas also opened up endless possibilities of using traditions such as the Puranas as evidence of scientific truth. It is possible to see that geomythology, which provides myths with geological meanings, is itself a product of the naturalization of antiquity that I am analyzing here. The history of shaligrams from the eighteenth to the nineteenth century explains how myths and geohistory became compatible with each other within this naturalistic imagination.

Avatars of Deep Time: Tortoises and Geomyths

Europeans came across tortoise/turtle myths in various parts of the world during the seventeenth century. Peter Young suggests that these myths originated in eastern Asia and then traveled to different parts of the world with human migration.[137] Myths of the world being carried on the back of a turtle were particularly common among Native American cultures. Edward Burnett Tylor, a believer in cultural evolutionism, sought to explain the presence of the "world-tortoise" mytheme across different cultures. He contended that the appearance of the same Puranic myths about the tortoise upholding the world in North America was a sign of cultural contact between the indigenous populations of Asia and the Americas.[138] Tylor interpreted these myths as representations of primitive ideas about the earth: the earth, which was believed to be flat, was the belly of the tortoise, and the sky was its shell.[139] Other scholars suggested that the tortoise was associated with these myths because its hard shell, strong legs, and slow movement represented protection, strength, and stability.[140] More recently, Young has explained the presence of these myths across cultures in terms of the "collective unconscious" by which various disparate traditions independently formulated similar ideas about tortoises and the earth based on the former's characteristic longevity, strength, endurance, and perseverance.[141]

Whatever the reasons for their presence in different cultural traditions, these myths have also been interpreted variously in different historical contexts. John Locke explained the Amerindian tortoise myths not as signs of wisdom but as evidence of ignorance, of ideas devoid of any clarity or basis.[142] In the eighteenth century, the tortoise myths were closely associated with biblical themes. Jones and other Orientalists interpreted them as reflections of the biblical deluge, as the tortoise avatar held up the world to save it from the flood. Amerindian tortoise myths too were given Hellenic and biblical interpretations by Europeans.[143] In the nineteenth century, these myths subsequently received a deep historical interpretation.

In the final section of this chapter, I explore how tortoise myths came to be associated with earth history as geology provided a new possibility in seeing the Puranas and Vedas as depictions not just of biblical history but also of geological deep time. We will examine the Puranic myth of the tortoise or the *kurma* (or *courma*) avatar of Vishnu, which came to be linked to geological ideas. Falconer linked these tortoise myths with geological deep time when he suggested that the *kurma* avatar of the Puranas was a real zoological creature that subsequently became extinct. There was a parallel tradition that explained these tortoise myths as nascent forms of geohistorical thinking and more specifically the tortoise avatar (along with other avatars) as allegorical representations of evolution, in the chain of transformation from invertebrates to humans. I will elaborate on the differences between the two positions later.

For the moment, it is useful to recognize the challenges that tortoises or turtles pose to the making of deep historical thinking. From the fossilized chelonias (tortoises) that the dentist and zoologist Thomas Bell found in London clay to the living ones in the Galapagos Islands that Darwin observed, tortoises have provided important clues about continuity as well as breaks in living and extinct fauna.[144] The American paleontologist Roland T. Bird described turtles as "one of God's best time travellers," as he believed they were around before the dinosaurs came to exist and survived the latter's extinction.[145] They were also, as indicated previously, associated with geomyths of creation.

Are these two imaginations, of evolution and geomyths, around tortoises related? Tortoises are significant because they link geomyths with modern evolutionary thinking. In this section I explore some of the ways these two parallel narratives around tortoises converged and how the geological explorations of the nineteenth century provided fresh allegorical readings of tortoise mythologies. Here I pick up the last strand of Falconer's use of Orientalism to assert the deep history of India. He used the tradition of mythological hermeneutics first developed by Jones and Wilford to invoke mythological animals from the Puranas as evidence of his explanation of the geological evolution of India.

The exploration of this theme takes us to the space between two paradigms of nineteenth-century intellectual traditions, Darwinian evolutionary ideas and Indology. On the one hand, it is necessary to extricate Falconer's Indian fossil research from its established place in the mainstream history of science, within Darwinian theory. Scholars consider Falconer's theory of stasis a critique of Darwin's theory of evolution.[146] It is essential to see Falconer's engagement with avatars and tortoise fossils as part of the broad narrative of Orientalist imagination of Indian antiquity, however. His idea of stasis was not just a critique or correction of evolutionary theory. As we saw in the previous chapter, Falconer, while explaining human evolution, drew a *longue durée* history of India, where the human past was extended to that of the fossilized animals that he and Cautley had found. On the other, Falconer did not belong to mainstream Indological scholarship. Although he was the first to link fossils with avatarism in the 1840s, his reference to the Puranas as the basis of his scientific theorization predated their late nineteenth-century Indological reading as proto-natural historical texts.

In 1834, while they were based in Saharanpur, Falconer and Cautley made their first major fossil discovery in the Timli Pass in the Himalayas, that of the gigantic tortoise.[147] The animal was believed to be more than 7 feet high, and its shell measured 18 1/2 feet in diameter.[148] They originally named it *Megalochelys sivalensis*, based on its size and place of discovery, but later named it *Colossochelys atlas*, both for its colossal size and as a reference to Atlas to denote the "the mythological tortoise that sustained the world, according to the systems of Indian cosmogony."[149] A few months after their discovery of the tortoise fossil, Falconer and Cautley found the fossil teeth and jaw of a primate in the Siwaliks that I referred to in chapter 2. Around this time, they also started some of the earliest paleontological deliberations on the links between apes and humans, which I explore in the next chapter.

In their article written in 1836, Falconer and Cautley sought to explain the connections between their discoveries of further primate bones, including facial and leg fragments, and the remains of the large tortoise.[150] They argued that the primate fossils, which closely resembled living species of monkeys, were found alongside fossils of crocodiles, which also resembled the living ones found in the Indian rivers. They referred to the discovery of the fossilized femur and shell of the tortoise, whose legs were the size of a rhinoceros's but otherwise was very similar to the much smaller living species of tortoises in India. The resemblances between fossilized and living animals suggested a stasis in the evolution of the Indian natural world, where living species resembled extinct ones. Based on this hypothesis, they suggested that the *Colossochelys* could be the mythological tortoise of the Indian Puranas: "As the Pterodactyle more than realized the most extravagant idea of the Winged Dragon, so does this huge Tortoise come up to the lofty conceptions of Hindoo mythology: and could we but recall the monsters of life, it were not difficult to imagine an Elephant supported on its back."[151] The implicit suggestion was that those who had composed the Puranas had based them on the giant tortoise.

This was their first and tentative reference to Indian mythology and a mythological interpretation of Indian fossils. It became a clearer argument once Falconer went to London in 1842. In 1844, he wrote from London to Cautley in India that he had put together the indications from Hindu mythology and the discovery of *Colossochelys* to advance "the opinion that the large Tortoise may have survived also, and only become extinct within the human period. *This is a most important matter in reference to the history of man.*"[152] He wrote a more detailed account in 1844 while in England.[153]

Falconer's assertions came at a time of renewed interest within Orientalism in the Puranas. The Sanskrit scholar Horace Hayman Wilson translated the Vishnu Purana into English in 1840.[154] The text described various episodes of Lord Vishnu and his different avatars. Wilson's *Vishnu Purana* became one of the original sources of subsequent deliberations on avatarism and Hindu evolutionary thinking.[155] It is not evident that he was aware of Wilson's translation, but Falconer was a regular visitor to the Royal Asiatic Society in London, which then, and for a long time (1837–60), was under Wilson's directorship. What is clear is that Falconer in the 1840s was one of the first scholars to present the Puranic idea of avatars as geohistorical concepts.

In 1844, Falconer gave several talks in Britain in which he referred repeatedly to the discovery of the *Colossochelys atlas* and narrated stories of Indian mythology. In these talks, he made the point that humans had existed in India from a very early period. The tortoise avatar of Indian myths referred to the *Colossochelys*, which the myths' composers originally saw as a living creature. To Falconer, the Puranas were therefore not allegorical texts. The tortoises in the Puranas were real zoological creatures, not symbolic Hindu representations.

In his talk at the Zoological Society of London, Falconer placed several large fragments of *Colossochelys* bones on the table and an illustration of the animal.[156] He started by referring to "the mythological tortoise that sustained the world, according to the systems of Indian cosmogony."[157] Before an audience consisting of British zoologists and paleontologists, he presented a narrative of zoological evolution using Indian mythology. He suggested that the *Colossochelys atlas* was associated with other major fossils found in the Siwaliks—the mastodon, rhinoceros, hippopotamus, horse, *Anoplotherium*, camel, giraffe, *Sivatherium*, and crocodiles. Several were "absolutely identical" with living species in India. The crocodile fossils were similar to those found in the Indian rivers, and the *Colossochelys* was similar to the *Emys tectum*, a common species of tortoise found all over India.[158] Falconer then asked his audience about the link between myth and reality: "Was this tortoise a mere creature of the imagination, or was the idea of it drawn from a reality, like the Colossochelys?" He referred to the presence of tortoise fables among Amerindians and went on to narrate in detail the Puranic stories of the *kurma* avatar of Vishnu: when the ocean was churned by means of the mountain Mundar placed on the back of the king of the tortoises with the serpent Asokee used for the churning rope. Vishnu was made to assume the form of the tortoise and sustain the world on his back. To a room full of scientists, he quoted, from Jones's sacred geographical accounts, some of the most fantastic mythological depictions of the tortoise: "The earth stands firm on thy immensely broad back, which grows larger from the callus occasioned by bearing that vast burden. O Cesava! assuming the body of a tortoise, be victorious! Oh! Hurry, Lord of the Universe!" He then delved into other references to the tortoise in Indian mythology, for example, in confrontation with the "bird-demigod" Garuda, who was also Vishnu's mount, where the tortoise was described as 80 miles long and the elephant 160 miles high.[159]

Falconer then provided geohistorical explanations to these myths. He first raised the question of the scientific credibility of these Indian myths: "Are we to consider the idea as a mere fiction of the imagination . . . or as founded on some justifying reality?" He asserted that although Greek and Persian mythologies presented more "fanciful and wild combinations" of known animals into "impossible forms," which even Cuvier had believed to be "merely the progeny of uncurbed imagination," Hindu myths presented an "an image of congruity." The paleontologist found the world of Siwalik fossils that he and Cautley had unearthed in the Himalayas in these myths:

> We have the elephant, then as at present, the largest of land animals, a fit supporter of the infant world; in the serpent Asokee, used at the churning of the ocean, we may trace a representative of the gigantic Indian python; and in the bird-god Garūda, with all his attributes, we may detect the gigantic crane of India (Ciconia gigantea) as supplying the origin. In like manner, the Colossochelys would supply a consistent representative of the tortoise that sustained the elephant and the world together.[160]

Based on "this congruity of ideas, [and] this harmony of representation," Falconer suggested that the Indian myths of the tortoise "as a symbol of strength" could not have been based on the small species currently found in India. They referred instead to the large *Colossochelys*, and humans in the Indian subcontinent who had composed these myths originally had lived at the time of the extinct tortoise. "The result at which we have arrived is, that there are fair grounds for entertaining the belief as probable that the Colossochelys Atlas may have lived down to an early period of the human epoch and become extinct since."[161] Without any direct geological evidence to suggest a link between the *Colossochelys* and the human period, he depended on "traditions connected with the cosmogenic speculations of almost all Eastern nations having reference to a tortoise of such gigantic size, as to be associated in their fabulous accounts with the elephant."[162] He then presented his main thesis, that humans had originated in the subcontinent at a remote age when this extinct species roamed the earth.

Contemporary naturalists in Europe were reestablishing links between fossils and mythological animals. Richard Owen, the conservator at the Hunterian Museum in London, referred to the pterodactyl as a flying reptile or dragon in his catalog published for the Great Exhibition in 1851. John McGowan-Hartmann explains that Owen and several other early nineteenth-century naturalists deliberately referred to fossils as mythological animals to introduce the idea of fossils to their audiences, to popularize them through something with which the audiences were familiar.[163] Falconer, while speaking to his audience in Britain around the same time as Owen, did something quite different. He used Indian mythological animals to explain scientific facts, to suggest a very different idea of myths: that these mythical creatures provided clues to paleontological research. Moreover, Falconer's talk at the Zoological Society was not a public lecture or a publication for a general audience, like the catalog of prehistoric animals Owen wrote for the Great Exhibition.[164] This was a lecture delivered to British zoologists and paleontologists, in front of whom Falconer presented these myths as evidence of his theory of deep time in India. In a single maneuver, Falconer provided a natural-historical interpretation of the Puranas and established the compatibility between Puranic myths and European deep history.

In his two subsequent lectures at the Royal Asiatic Society in London, referred to earlier, he repeated the point about the early origin of humans in the subcontinent and suggested that some of the Siwalik fossils that he and others had collected carried evidence of human life in India at "very remote times." He gave the example of the *Colossochelys atlas* and suggested that it fit the representations in Hindu and Pythagorean mythology of the tortoise supporting the elephant. Throughout his first talk at the Royal Asiatic Society, he referred repeatedly to the Hindu myths suggesting that humans lived in India at the time of the extinct tortoise.[165] In his second lecture at the society, Falconer elaborated on the lack of evolutionary change in Indian fauna, extant or fossilized, and asserted that every geological epoch contributed to the formation of "one comprehensive fauna" in ancient India.[166]

Scientists in Britain did not directly engage with Falconer's analytical and narrative strategy of using Indian mythology as scientific evidence or with the question of the presence of early humans at the time of the extinct Siwalik fossils. In their discussions, they ignored the mythological content of Falconer's talks. Lyell, who was present at the Royal Asiatic Society meetings, did not refer to it in his response and provided a climatic explanation for the extinction of the Siwalik fauna. He pointed out that this was the first collection of vertebrate fossils "procured from a country where the climate may be presumed to be as hot now, as at the period when the fossil animals flourished." He suggested that the similarities that Falconer had found between extinct and living species could be due to "considerable similarity of temperature in ancient and modern times" in India. He explained the extinction of the Siwalik fossils with glacial movement. The drifting ice and debris from the poles to lower latitudes could have destroyed numerous species in India. He also suggested that the Siwalik strata could belong to the older Pliocene or even the Miocene period, much earlier than Falconer had presumed.[167]

Although Lyell did not engage with Falconer's mythological hermeneutics, he was aware of the significance of Vishnu's avatars to geology. In the seventh edition of his *Principles of Geology*, published around the same time of Falconer's talks, Lyell referred to Hindu creationism, the Puranic narratives of repeated destruction and revivification of "organic beings," the recurring submersions of the land beneath a "universal" ocean with Vishnu successively assuming the fish, tortoise, and boar avatars, as evidence of Hindu prehistoric knowledge. Deriving from Colebrooke's and Wilson's work, Lyell suggested that since the remains of the marine fossils on the rocks and stones were "so abundant" in the Himalayas, the Brahmins who wrote about the avatars derived these ideas from these fossils. The suggestion, therefore, was that these texts contained early paleontological ideas.[168] He similarly believed that the priests of Egypt and the Greek historian Herodotus were aware of the marine fossils of the Nile.[169]

Lyell and Falconer represented two different interpretations of Indian myths and their geohistorical connotations in the mid-nineteenth century. For Lyell, the significance of Vishnu's avatars to geology was purely antiquarian; he saw the Puranas as evidence of embryonic Hindu geological thinking. For Falconer, however, they had scientific and empirical value: they provided the missing interpretative links in contemporary geohistorical knowledge. In other words, Lyell viewed the authors of the Puranic texts as primitive geologists who saw the tortoise as a fossil. Falconer suggested that the creators of the myths saw the giant tortoise as a living animal, thus using the texts as evidence of his paleontological theory of speciation. In doing so, he treated the *kurma* avatar as part of his modern paleontological investigation. Rather than suggesting that the composers of these texts were primitive geologists, he believed that myths such as these could provide missing geological clues.

Methodologically, his work was reminiscent of the comparative research adopted by Georges Cuvier in 1804 while studying a mummified ibis, a bird sacred to the ancient Egyptians. While examining the remains sent to him from Egypt following Napoleon's invasion, Cuvier depended on not only the analysis of bones and plumage but also descriptions of the bird by ancient writers such as Herodotus and Plutarch. He insisted that the mummified remains of the bird were identical to a modern species of curlew. Cuvier believed that studies of fossils would remain inaccurate without such comparative research.[170] Falconer's choice of these mythological texts for such comparative research is unique, particularly in the middle of the nineteenth century, when the Puranas had become antiquarian points of reference to modern scholarship. Herodotus and Plutarch, on the contrary, were heralded as some of the founders of the modern and objective historical method. Falconer's geohistorical frame too was distinct from that of Cuvier. His research on the Siwalik fossils led him to reject Cuvier's catastrophism along with Buckland's diluvianism. He believed that the continuum between fossils and living fauna were signs of "tranquillity" on the "great Indian plains" rather than any "violent" event that those theories suggested.[171]

Lyell's view that ancient Hindus were early evolutionary thinkers was more compatible with the mainstream interpretation of avataric evolutionism of the 1870s. It was also part of a wider intellectual tradition in the late nineteenth century that regarded myths as proto-scientific ideas. Scholars such as Tylor similarly referred to Polynesian, Amerindian, Australian Aboriginal, and Central Asian myths as primitive reflections of deep time.[172] Such views of myths carrying nascent geological insights also conform to the propositions of the modern scholarship of geomythology.

In the early twentieth century, Pavgee presented a geohistorical reading of the *kurma* avatar. He incorporated late nineteenth-century avataric evolutionary ideas to suggest Hindu myths as evidence of early geological ideas. In his pursuit of a true geohistorical tradition of the Hindus, he placed more faith in the earlier Vedic texts than in the later Puranas, which he believed had subsequently incorporated several allegorical interpretations. In doing so, he established Hinduism itself as an essentially scientific tradition that contained true interpretations of nature. He regarded the Hindu sages who composed the Vedas as "fathers of geology" and the Vedic literature as geological texts, with accurate references to Paleozoic and Mesozoic animals, such as the tortoise. These Vedic ideas received mythological and avataric interpretations in the later Puranas, when according to Pavgee, religious ideas were introduced into "everything Hindu."[173] As he treated the Vedas as essentially geological texts, his interpretations were more literal. According to him, the earth in the Vedic age was still cooling down. The tortoise, depicted as saving the earth from the rising water in the Puranas, was in fact originally represented in the Vedas as an example of reptilian life in an earth that was still in a "liquid state."[174] He suggested that each subsequent avatar in fact originally appeared in the Vedas as a reference to the emergence and evolution of various stages of life in different geological epochs. In establishing the mythological tortoise as a geological species, Pavgee arrived at Falconer's conclusions, although he did not conform to the latter's theoretical rationale. While Falconer had suggested that some of the earliest humans on earth, who referred to the giant tortoise, composed the avatar myth, Pavgee believed that the Vedic scholars (who were aware of geological epochs) referred to the tortoise as a typical reptilian animal of the prehistoric liquid state of the earth. In both depictions, however, the *kurma* avatar was naturalized as a species of deep time.

What happened to the fossil itself, the *Colossochelys atlas*? In subsequent years, Lyell established that the giant tortoise lived only up to the Miocene era, undermining Falconer's theory about the tortoise having survived into the human period.[175] From the 1850s, following his encounter with Boucher de Perthes, Falconer was preoccupied solely with the question of human origin and the excavations at Brixham Cave. The *Colossochelys* receded from his interest. It seems that fragments of the shell were brought back to India and are now kept, along with later ones, in different museums.[176] Richard Lydekker subsequently wrote on it and other large fossilized tortoises found in Asia. He suggested changing the name *Colossochelys atlas* to *Testudo atlas*, as the animal belonged to the genus *Testudo*.[177] In the early twentieth century, paleontologists became interested in it for both its unusual size and its connections with tortoise myths. The American paleontologist Barnum Brown, famous for his discovery of the *Tyrannosaurus rex*, took a keen interest in the giant tortoise fossil, attracted by its size. He led an expedition to the Siwaliks and northern India in 1923 and collected the "the first and only complete shell of its kind known," rather than the fragments collected by Falconer and others in the nineteenth century.[178] Brown and his colleagues gathered all the pieces and loaded them, with some "military coercion," on the back of a seemingly disinclined camel. The sight of the large camel struggling with the load of the massive tortoise on its back as it walked down the hills reminded Brown of "many a tortoise legend" that referred to its "gigantic size." He then recounted almost verbatim Jones's and Falconer's writings about the tortoise avatar of Vishnu, lifting the "infant world . . . on the back of an elephant." Brown noted that a reluctant camel was now carrying the extinct tortoise on its back.[179]

At the American Museum of Natural History in New York, scientists worked hard to piece the shell of the tortoise together. The fully reconstructed shell was 7 feet, 4 inches in length over the curve, 5 feet wide, and almost 3 feet in height.[180] To compare its size, Brown placed the shell of a Galapagos tortoise underneath it (fig. 3.3).

Brown estimated that the fossilized tortoise he discovered weighed a ton when alive and was 300 or 400 years old when it died. Resorting once again to myths, now the Aesop's fable about the hare and the tortoise, he declared that the "Atlas [*Colossochelys*] has won a race, and posterity has the privilege of viewing a masterly finish."[181] The fossil of the *Colossochelys atlas* that Brown collected in India is now displayed at the museum as *Geochelone atlas* (fig. 3.4). It is now an avatar of deep time.

Fig. 3.3. Otto Falkenbach at work on giant tortoise fossil. Courtesy American Museum of Natural History Library, image 22444.

Conclusion: Naturalizing Sacred Geography

Nature and geography came to be the core themes of evolutionary thinking in the nineteenth century. Scientists explained transformations and differences in extant and extinct nature through the then-novel concept of evolution and extinction. However, these explanations were based not just on the observation of nature itself but also on multiple cultural conceptions of nature. From the eighteenth century, Europeans interpreted the natural history of various parts of the world through both biblical and local mythological motifs. Therefore, while the idea of evolution itself unfolded largely within secular premises in the late nineteenth century, the cultural and sacred ideas embedded within this wider experience of the history of nature played important roles. In the process, sacred geographical themes received a new lease on life in deep history.

Historians have analyzed how geology was shaped by myths and how geologists found clues about the history of the earth from various myths. This

Fig. 3.4. Barnum Brown with his tortoise fossil. Courtesy American Museum of Natural History Library, image 22491.

chapter (and chapter 1) has analyzed how the character of myths and sacred geographies themselves transformed and acquired deep historical undertones. There have been attempts at mapping mythologies and sacred geographies since the medieval and early modern period. Al-Biruni, the eleventh-century Arabic scholar, polymath, and preeminent exponent of medieval enlightenment, for example, sought, often in vain, to ascertain the correct cartographical dimensions of Indian mythological geographies from Sanskrit texts. The measurements of Meru, references to which he had come across in the Puranas, appeared to him to be grossly "extravagant."[182] Similarly, there had been several attempts at mapping paradise since antiquity.[183] From the eighteenth century, there were also efforts at aligning global geography with biblical notions of time and nature. During this time, as evident in the works of Jones, Maurice, and Wilford, mythical, textual, and physical ideas of Indian nature came to coexist. In their work, therefore, the celestial and the terrestrial nature of the Orient cohabited.

In the nineteenth century, these redefinitions of sacred geographies were placed alongside emerging ideas of the geological history of nature. In India, unlike in Europe, for the deep history of the earth to be established, its sacred history did not have to recede. The discovery of deep time and the geological mapping of colonial landscapes in the nineteenth century provided a deeper naturalistic scope to Orientalist readings of myths. Consequently, the Saraswati, Aryavarta, Puranas, and the avatars appeared as profound entities of Hinduism and have been inscribed into the deep Hindu psyche and landscape of India.

Remnants of the Race

Geology and the Naturalization of Human Antiquity

As the British army officer and antiquarian John Henry Rivett-Carnac traveled from the Yamuna-Gangetic Doab toward central India, he watched the landscape change dramatically. There were no longer the "monotonous" northern alluvial plains dotted by trees and groves. Within a few miles south of Allahabad, the city situated at the confluence of the two great northern rivers, the Ganga and Yamuna, rocks, hills, and jungles suddenly appeared, leading to a picturesque terrain of valleys and highlands.[1] This was the district of Banda, which formed the border between the northern plains and the densely forested and hilly regions of central India, an entry point into a different country and a distinct antiquity. Rivett-Carnac was captivated by this landscape, which it seemed "civilization ha[d] not even yet fully penetrated or robbed of its many sylvan attractions."[2] He noticed that the mythological references too changed with the landscape and were now tinged with aspects of primitivism. He had learned about this region being the frontier between civilization and savagery from Indian epics such as the *Ramayana*, according to which, Rivett-Carnac reminisced, Lord Rama, after having resigned from his kingdom in Ayodhya in the northern plains, resided there in Banda along with his wife Sita and his brother Lakshman. The myth described these picturesque landscapes where Ravana, the king of the savage demons, appeared and abducted Sita and where Rama formed his army of monkeys to retaliate and rescue his wife. Rivett-Carnac believed these mythological monkeys were, in reality, the "the wild tribes inhabiting these tracts, who were probably armed with the stone hatchets and the stone clubs." These people and their stone tools were the subjects of his antiquarian inquiry, as he found "semicivilized tribes" living in these hills and prehistoric stone implements hidden beneath the rocks nearby. He referred to ancient Hindu records that contained accounts of "these wild men of the woods, and the ancient stone carvings" similar to those he had found.[3] He was seeking to identify the aboriginal tribes living in the area who presumably were familiar with these prehistoric tools.

Rivett-Carnac collapsed the deep prehistoric past with the colonial present by employing mythology, paleontology, and anthropology simultaneously, all underpinned by references to the terrain, to rehearse an apparently simple assumption that the aboriginal populations of this seemingly primal landscape were remnants of prehistoric humans. He was not alone in doing so, nor were Banda and central India unique sites for making such assumptions. In the nineteenth century, there were common references to aboriginal populations in India, Australia, South America, and Africa as remnants of prehistoric humans. For example, the geologist and ethnologist George William Stow's find of prehistoric rock paintings in the caves near Queenstown in South Africa in the mid-nineteenth century inspired his ethnological interest in contemporary Bushmen[4] art, as he believed that both were part of the same cultural heritage.[5]

Why did tribal populations appear to be remnants of prehistoric humans? This seemingly simple yet deeply problematic association of living populations with those of thousands of years ago was made in the nineteenth century. In earlier accounts, both in Europe and beyond, tribal or marginal populations were compared to "barbarians" and "savages" but not to prehistoric humans, because the idea of human prehistory itself had not emerged before the 1830s. Tournal and Boucher de Perthes proposed the deep geological antiquity of humans in the 1830s and 1840s, based on their paleontological discoveries of human remains and stone tools, respectively. The publication of Daniel Wilson's *Prehistoric Man* (1862) and John Lubbock's *Prehistoric Times* (1865) marked the beginning of British deliberations on prehistoric archaeology, along with the appearance of the term "prehistory" in English.[6] Once humans were assigned to prehistoric times, the study of aboriginal populations became entangled with the study of prehistoric humans. The geological and archaeological explorations of human origin took place around the same time that European ethnologists encountered different aboriginal populations in Asia, Africa, the Pacific Islands, and South America. The objectives of their inquiries thereby merged. The main quest for Victorian anthropologists, as summed up at the 1874 Anthropological Exhibition in Bethnal Green, in London, was to explore "to what extent the modern savage actually represents primeval man."[7]

Answering the question, why the aboriginal populations appeared to be remnants of prehistoric humans also opens the possibility of understanding how habitat, nature, and landscape became integral to discussions of both human aboriginality and tribal history. References to the geology and ecology of the immediate situations of aboriginal or tribal populations were often foundational to the inquiries into their histories. The French paleontologist Marcellin Boule found "everything" about the Australian landscape "peculiarly" prehistoric. Its vegetation, coral beds, marine life, and reptiles all reflected nature in its embryonic form within which he situated its native humans and their manifest primitivism: "The indigenous human populations *likewise* belong to one of the most primitive modern races."[8] For geologists and ethnologists, certain regions in the colonies and the tropics, whether remote islands or densely forested hills, appeared to be sites of prehistory, where human populations not only appeared aboriginal but also seemed to have remained relatively unchanged since prehistoric times.

Evolution has been one of the main frames through which to explain the connections between primitivism and aboriginality. On the one hand, Falconer's theory of speciation, which suggested that certain species in some parts of the world did not follow the evolutionary pattern and had reached stasis, could partially explain why some human races had remained prehistoric. On the other, historians have identified Charles Darwin's *Beagle* voyage to South America and the Pacific in the 1830s, and his observations of indigenous populations conducted along with his ship captain, Robert FitzRoy, as a significant moment when aboriginal races became part of evolutionary thinking. Janet Browne has explored FitzRoy and Darwin's observations and treatment of the "Fuegians" (the people of eastern Tierra del Fuego, in South America). She sees the contrasts they drew between these people and "civilized humanity" as part of the concern for the former's mental and moral "improvement" and "progress," with the underlying belief in the basic commonality of all humans that preoccupied British missionaries, biologists, and anthropologists in the nineteenth century.[9] Peter Bowler sees Darwin's attitude toward Fuegians and Australian Aboriginals as a combination of the anthropologist Edward B. Tylor's ideas of cultural evolution and Darwin's own emerging concepts of biological evolution. Darwin assigned these aboriginal people at a different level of racial and cultural evolution compared to presumably the nonindigenous people. Drawing subsequently from Lubbock's and Tylor's observations of "Stone Age" humans, Bowler conflates observations of primitivism with observations of prehistoric humans within the same evolutionary frame of human progress. He explains that Lubbock and Tylor saw aboriginal people resembling prehistoric races as part of human evolution.[10]

There was yet another frame that explains the transition from the primitive to the prehistoric. Ethnologists and geologists were not just interested in racial evolution or progress from primitivism to modernity. They also referred to the seemingly prehistoric existence of these humans in front of them, along with the artifacts of prehistory, such as the flint stones, fossils, and *celts*[11] that were unearthed in the same region and around the same time. This facility of situating these living populations within the prehistory of the earth was particularly available to geologists and ethnologists based in India, South Africa, Australia, and even parts of Europe. British scientists, who often saw portraits and ethnological descriptions of these people and received their tools and the fossilized artifacts removed from their immediate geological settings, tended to view them in isolation. Although this distance allowed them to put these people within a grand scale of global human evolution, it did not always enable them to place them in their immediate geological settings.[12]

While observing certain races as "prehistoric," anthropologists and geologists in India did not necessarily see them within any evolutionary pattern. Nor were they always seeking to improve their souls, civilize them, or convert them to Christianity. Even their reference to these people as apes, as we shall see, came not purely from an evolutionary point of view. For them, tribal primitivism was more geological than evolutionary; it was in their *state of being*, as fossilized relics in the ostensibly pristine and prehistoric landscape that surrounded them. They observed these people as geological specimens. Therefore, their studies had strong parallels with geological observations, and thereby the question of why humanity became part of earth history needs to be explored here.

The questions about the links between prehistory and aboriginality might not appear significant from certain perspectives. Alison Bashford has observed that in Australia, the collapse of prehistory into the Aboriginal present is political. On the one hand, the idea of a "continuous" culture of 50,000 years is an important part of Australian Aboriginal identity and their rights to the land. On the other, in the Aboriginal notion of time, "chronologies of before and after make little sense, and don't need to: 'ancient and modern,' 'past and present,' 'deep and shallow,' are adjacent temporalities."[13] Ajay Skaria makes a similar point in his study of the Bhil tribes in the Dangs region of western India. According to him, their *goths*, or stories of their ancestors, contain notions of time and truth that are inconceivable within Western or modern chronological definitions of history or prehistory. In their tradition, the precolonial (*moglai*), which is synonymous in the Bhil world with freedom, is "coeval" with various themes of time.[14] Therefore, the frames of history and prehistory could be modernist impositions on aboriginal notions of time and past. While these caveats are true, it remains to be explored whether, on the one hand, European ideas of prehistory, with their inherent naturalism, influenced tribal and aboriginal naturalism and origin myths, and vice versa. It is also important to understand why environment and the deep history of the landscape appear so critical to understanding tribal aboriginality and indigeneity. These inquiries are relevant because European encounters with aboriginal notions of time and discoveries of their own geological prehistory took place simultaneously. According to Tom Griffiths, for some Aboriginal populations of Australia, deep time also carries "all the baggage of nineteenth-century evolutionary thought": memories of racial prejudice, social Darwinism, and derogatory representation of them as a "primitive" race.[15]

Aboriginality and indigeneity have been critical themes in postcolonial literature not just in settler colonies such as Australia, South Africa, or Canada but also in South Asia. Scholars have often defined aboriginality in sharp contrast to the colonial and postcolonial appropriation of land, forests, mines, and the associated traditions and rights of tribal communities.[16] Tribal and aboriginal movements against state policies and environmental destruction have often shaped these academic engagements.[17] However, these discussions, at least in South Asia, do not critically engage with the conceptual fluidity between aboriginality and prehistory. This is mainly because researchers have taken two distinct approaches. On the one hand, historians have examined aboriginality, indigeneity, and primitivism mainly from ethnological and anthropological perspectives without problematizing their synergies with geology and prehistory.[18] On the other, scholars have studied human prehistory in India from predominantly paleontological or paleoanthropological perspectives. Kennedy, in particular, has provided extensive paleoanthropological accounts of human origin in the Indian subcontinent in which ethnological questions have not been central.[19]

Peter Pels recognized the fluidity between textual and physiological studies of Indian aboriginality, through the works of two nineteenth-century scholars, Brain H. Hodgson and Max Müller. However, Pels does not really explain this transition from texts to bodies, as the two individuals he studies did not conduct any physiological ethnology in India and depended almost entirely on textual sources.[20] Therefore, the overlapping of different disciplinary investigations, of texts, relics, and bodies remains to be examined. Sumit Guha has examined the scholarship, which assumes the synonymy between tribal populations and aboriginality to suggest a genetic continuity between the tribes and the prehistoric human populations.[21] Guha has analyzed the archaeological evidence, which shows "continuous interaction involving both conflict and cooperation" between different types of communities; thus, they were neither fixed to any ordained location, nor were they confined to a certain changeless time warp.[22]

While Guha has corrected the simplistic assumptions behind the aboriginal-tribal continuum, this chapter extends that examination to the assumptions made collectively by paleontologists, philologists, missionaries, and ethnologists in the nineteenth century that tribes were remnants of prehistoric human populations. In addressing these themes, I scrutinize how deep history superseded other forms of existing ethnological and philological deliberations about human antiquity. These themes inform the wider question that drives the narrative of this chapter, how humans became naturalized entities.

In doing so, I return to the two core arguments of this book, of seeing the deep past in the present and of the links between the deep past and the present that were formed across different disciplines, in this case paleontology, anthropology, and zoology. This synergy across the various disciplines provided the emerging discipline of anthropology with greater scientific validity and in turn, gave the entire proposition of finding prehistoric humans among living populations a firm intellectual foundation. The anthropological exhibition at Bethnal Green in London organized by the South Kensington Museum in 1874 displayed the convergence of zoology, anthropology, and paleontology in the study of humanity—of apes, tribes, and prehistoric humans. The exhibition catalog stressed the overlapping nature of the assignment: "What the palæontologist does for zoology, the pre-historian does for anthropology. What the study of zoology does towards explaining the structures of extinct species, the study of existing savages does towards enabling us to realise the condition of primeval man."[23]

This disciplinary consilience entailed simultaneous journeys into the deep past and the present: on the one hand were the investigations into fossilized prehistoric human skulls and bones, stone implements, and burial remains; on the other were the ethnological studies of living tribal populations as aboriginal inhabitants. These were substantiated by zoological and racial studies among tribal populations as descendants of primates. The question lies in understanding how the different inquiries involving paleontology, zoology, and ethnology were fused and how, for example, like Rivett-Carnac, nineteenth-century scholars came to assume that the tribal populations they encountered were familiar with or had even used the prehistoric tools that they found in the neighboring regions.

This is not the first or only instance when humans were naturalized. The Linnaean project of placing humans in the natural order of classification in the eighteenth century provided the important clue to seeing humans and particularly races as biological species and as part of natural history. The climatic, biological, and physiological conceptions that dominated the eighteenth- and nineteenth-century debates on the human race and racial hierarchy, which Stephen J. Gould described as the "geometer of race," was an important instance when human beings were seen in biological terms.[24] These are not the main subjects of this inquiry. The Linnaean classificatory system or the climatic, phrenological, and biological determinism of race did not refer to the geological configuration of human antiquity that I examine here. At the same time, as I will explore, the search for the deep history of human antiquity crossed paths with the biological idea of race in identifying primates as human ancestors.

There are five interrelated themes here. First, the discovery of prehistoric humanity, which took place in the nineteenth century, with paleontological investigations. Second, the simultaneous inquiries into ethnological aboriginality and its entanglement with prehistory. Third, the aboriginal ideas of their own prehistory, which I will explore more in the next chapter. Fourth, the connections established in India between aboriginality and primates both within and beyond Darwinian evolutionary ideas, which provided another link between tribal antiquity and the history of nature. Finally, as I will establish in the case of India, the aboriginal populations were observed not just within an evolutionary scheme of moral and physical improvement but as prehistoric specimens, as part of the geological landscape, through the analysis of prehistoric tools.

Aboriginality as Prehistory

The search for human antiquity in India coincided with one of the main sites of economic and political interests of the colonial state in the nineteenth century. This was the vast and densely forested region known in colonial records as the Central Provinces (formed in 1861), on the fringes of which Rivett-Carnac had found himself.[25] The colonial exploitation of natural resources and the search for human prehistory in the same region integrated these natural and historical investigations. From the late eighteenth century, the British had sent several expeditions prospecting its vast natural resources. Geologists found rich seams of coal, mines of diamonds, and fertile tracts of black soil ideal for growing cotton. They also found tribal populations living in isolated hills and forests who they believed were the relics of prehistory. These populations, in turn, inspired missionary activity in the region. As a result, central India in the nineteenth century became a fertile ground for British economic, geological, ethnographical, and antiquarian encounters.

The British occupied Mysore in the south by the end of the eighteenth century, which fueled their ambitions toward central and western India and led to conflicts with the Marathas. Warfare with the Marathas started at the end of the century in which Nagpur, one of the capitals of the Maratha rulers of the region, occupied a central place. With the final defeat of the Marathas by British forces in 1818, large tracts of central India, including the Deccan, were annexed to the territories of the East India Company.

Central India presented British explorers and scholars with very different propositions than had the Himalayas or the Gangetic Plain of northern India. While the latter invoked ideas of the Indo-Aryan race, believed to have descended from the lofty snow-capped mountains, the hills, forests, plateaus, prehistoric relics, fossils, and tribes of central India reflected ideas of tribal aboriginality. They provoked different mythological anecdotes, those of Aryan conflict with aboriginal tribes. The region presented prospects of a deep past that were poised to overwrite the Orientalist, Hindu, and Islamic antiquity of India. The search for the aboriginal races of India developed in response to the British encounters with caste Hindus and Muslims there, who they believed were immigrants and settlers in this region. Central India was a triumph of the "field," of nineteenth-century sciences such as geology, anthropology, and archaeology. The simultaneous geological and ethnological explorations shaped the imagination of both this landscape and its aboriginal inhabitants as primal and prehistoric.

Colonial geologists arrived there with large entourages. Valentine Ball, while prospecting for coal and diamonds, was accompanied by a native doctor, a headman who was in charge of the Indian staff, and several *chaprassies* (orderlies) who performed duties such as pitching tents, acting as messengers or night watchers, and obtaining supplies from the villages. There was also a valet who managed the internal arrangements of the tent, a cook, a cook's assistant, the table attendant, as well as the washerman, *bhisti* (water provider), and sweeper. He also had elephants and bullocks as beasts of burden.[26] The entourage often recruited local people, which, as we shall see, became the basis of some of Ball's cultural and ethnological encounters in the region.

Missionaries came to central India around the 1840s with more modest resources. Donald McLeod, a colonial administrator of the Central Provinces and a deeply Christian man, was instrumental in establishing missionary activities among the tribal populations of the region.[27] He believed that since the tribes of these regions were similar to the aboriginal races of Africa, America, and the Pacific Islands, where Christian missions had achieved great successes, missionary activity among them would be similarly fruitful.[28] As no English missionary society was initially willing to take up this idea, he applied to

Pastor Johannes Gossner of Berlin, who sent a small group of German missionaries to Jabalpur in 1841 to work among the tribes. They worked under the superintendence of the Reverend J. Loesch, a Lutheran minister of the Basel Missionary Society, who had previously worked in south India. They served in the hilly regions, particularly near the source of the Narmada River, at Amarkantak, a sacred Hindu pilgrimage site known as the "Waters of Immortality."[29] Just beyond this site, in the forested hills, lived the Gonds, the largest tribal population in the region. The Germans started their first missionary activities among them. They also started to compile a lexicon of Gond words and noted that many were of Kannada or Tamil (the two major languages of southern India) origin.[30]

Stephen Hislop, a Scottish missionary who played the most influential role both in his missionary activities and his chronicling of tribal genealogies, came to Nagpur in 1845 as a volunteer of the Free Church of Scotland to set up a mission and schools.[31] He soon learned Marathi and in 1849 established a girl's school in Nagpur. The school at Kampti for British military children was converted from a cantonment school. He established another school at Sitabaldi in Nagpur in 1849.[32] In 1847 Robert Hunter, a fellow Scottish missionary, joined him. During their long walks together, they searched for fossils near Nagpur, studied tribal culture and genealogies, and speculated about the tribes' prehistory. As they preached the Bible among them, they also sought to identify their "pure" racial and cultural heritage, untainted by Hindu influence.

As Sangeeta Dasgupta has shown, early missionary discussions of tribal aboriginality in India were nonlinear and negotiated diverse categories such as heathens, pagans, and savages, depending on the context of their exchanges with different tribes. These categories were gradually subsumed within a more common theme of primitivism by the middle of the nineteenth century.[33] The missionaries and ethnologists also found references to tribal populations as *dasas* (slaves), *dasyus* (enemies), and *asuras* (demons) in Sanskrit texts,[34] which informed their assumption that these tribes were the aboriginal inhabitants of India who had been subsequently enslaved by successive waves of conquerors and settlers, starting with the Aryan races. Subsequent anthropological studies of their physical features and their cultural and religious practices, all of which appeared to be at odds with Brahmanical practices, confirmed these ideas. Christian missionaries, in particular, reinforced their non-Sanskritic roots by suggesting that the tribes had been co-opted by (and needed to be saved from) the Hindu caste system.

The question of aboriginality developed within a peculiarly colonial milieu posited against the notion of the settler. In different colonial settings, these settlers were either Europeans themselves or other local populations. In Australia, the indigenous populations were imagined as the aboriginal and in situ with the landscape that itself appeared primitive, posited against the white, racially "superior," and "modern" European settlers. In South Africa, the San people, or the "Bushmen" (which in itself is a natural-historical term) were regarded as aboriginal to the land, against the Bantu people or more specifically the Xhosa (Nguni), with whom the British were involved in several conflicts from the late eighteenth century and whom they regarded as relatively recent settlers in the region. In India, aboriginality was framed at various intellectual sites and often at abstract levels in response to different inquiries into the history of human origins. The only commonality was that they were all situated around the question of non-Aryanism. Discussions about primeval humans in India emerged from theological concerns about the late eighteenth-century ideas of Hinduism's Judeo-Christian roots: Jones and other Orientalists believed that Indians were the descendants of Ham. In contrast, the ethnological investigations in the nineteenth century were undertaken to establish the aboriginal races living in India before the arrival of the Aryans.

The deep past of humanity collapsed into its present because the ethnologists and missionaries in India were seeking to resolve a contemporary problem of placing existing populations in their "appropriate" order of temporality. The search for a racial lineage that preceded the Aryans and Hindus led to seeing tribes as autochthones of India. The missionaries drew from and contributed to an emerging theme of aboriginality in India that challenged theories of Aryanism: "Dravidianism." While Indo-Aryanism suggested the Aryans' eastern origins, Dravidianism proposed that the Dravidian races were the aboriginal inhabitants of India. As German and British missionaries noticed similarities between the tribal and the Dravidian languages, they pronounced the idea that the clue to Indian aboriginality lay in understanding Dravidianism. This stress on an aboriginal Dravidian culture was important to the European understanding of tribal culture and heritage.[35] By the late nineteenth century, many scholars pointed to Dravidian culture as the repository of aboriginal cultures of India. Robert Caldwell was the first to claim that Dravidian peoples populated India prior to the Aryan invasion, having themselves originally entered the subcontinent from the northwest. Caldwell came to Madras in 1838 with the London Missionary Society and learned

Tamil and Telugu as part of his role. His assertion of Dravidian cultures as fundamentally distinct from Aryan ones derived from his missionary urge to find an alternative to Brahmanical cultural and racial heritage in India.

Based purely on phonetics and philology, Caldwell suggested that the Dravidian languages of southern India, such as Tamil, Telugu, Malayalam, and Kannada, were different from those derived from Sanskrit, and therefore these Dravidian cultures, societies, and polities had existed prior to the arrival of the Brahmins in the south. He then extended this into a broader thesis that the Dravidians were of Scythian (the nomadic Eurasian groups) origin and were closely related to Finnish or Ugrian races, which were at some point driven south either by the Aryans or by a race of Scythian intermittent invaders.[36] Gustav Oppert similarly claimed that central Indian tribes constituted "the first layer of the ancient Dravidian deposit."[37] Oppert was the professor of Sanskrit and comparative philology at Presidency College in Madras, and for him, language was the key to tracing the "continuity of descent from the same stock in tribes seemingly widely different"; the Gauda-Dravidians were the "Bharatas," or original inhabitants, of India.[38] Oppert suggested, like Caldwell, that the original inhabitants of India, or the "Gauda-Dravidian tribes," were identifiable with the race dispersed across Asia and Europe termed "Finnish-Ugrian, or Turanian."[39] The ethnologist-historian John Briggs, in his lectures on the aboriginal race of India, emphasized that the aboriginals throughout India came from a common racial stock.[40] Trautmann makes the point that by the mid-nineteenth century, this hypothesis of the Dravidian heritage of Indian tribal aboriginality had become "an article of faith."[41]

In the course of these textual philological investigations, two ideas became entrenched. First, the tribal populations of India were racially and culturally distinct from the Hindus. Second, they were the aboriginal races of the subcontinent. Missionaries played a significant role in originally repositing this sense of aboriginality in Indian tribes. Through the study of tribal languages, they sought to identify the deep historical roots of these populations.[42] In central India, Hislop and his fellow missionaries looked for a human antiquity that preceded Hinduism in India. This necessitated a rejection of the Puranic traditions and avataric imageries that were formative in the discovery of Hinduism by the early Orientalists.[43] According to David Arnold, the Evangelical attack on Hinduism from the early nineteenth century severed any possibility of appreciating Hindu imaginations of Oriental nature. The Hindu idols, gods, and iconography, which had opened up different possibilities at an earlier age, now appeared "obscene."[44] During a brief visit to Edinburgh in

1859, Hislop gave a lecture on the avatars of Vishnu, each of which, according to him, had "indulged in all manners of lasciviousness," and concluded, "the adherents of this religion are sunk in depravity and misery."[45] In contrast, the tribal people, who did not worship these Hindu deities, were noble savages, "honest and truthful, intelligent, energetic, and manly."[46] Hislop rejected the proclaimed mythic Rajput lineage of the Gonds, suggesting that they would "resent, with no small vehemence, imputation of belonging to any Hindu community."[47]

Armed with their knowledge of tribal languages, Hislop and other missionaries and philologists presented a much purer ethnic and cultural lineage of tribes as the aboriginal races of India. The study of tribal languages rejected the idea that Sanskrit was "the parent of the principal languages of India."[48] Rather, missionaries and ethnologists believed that "rude simple aborigines" of India spoke a "simple, homogenous dialect" before the arrival of the Aryans.[49] This emerging nineteenth-century philological tradition based on Dravidian and tribal languages circumscribed Jones's and Wilford's late eighteenth-century Orientalist philological practice, which drew from classical languages such as Sanskrit, Persian, and Arabic.

The philological deliberations on tribal languages gave rise to theories of tribal deep history that were defined in terms of the "pre-Aryan" history of India. Tribal languages such as Gondi were believed to belong to that simple primitive and original language of India. Missionaries such as J. G. Driberg (who later published a grammar and vocabulary of the Gondi language[50]) and W. Taylor, who served in the same region, believed that in the remotest mountains, Gonds spoke a language of "the greatest simplicity and purity."[51] Hislop studied Gondi and believed that all these languages and the tribes who spoke them originated in the mountains of northeast India, and then in Burma and as far as Malacca. He believed that two sets of migrations, one from the northwest and the other from the northeast, coincided somewhere in central India. The migrating tribes met the aboriginals of India, who eventually formed the Indian tribal population.[52]

These studies also suggested that all the aboriginal tribal populations of the subcontinent, whether the Karens and Shans of Burma or the Kols and Gonds of central India, belonged to the same racial spectrum.[53] The naturalist Brian H. Hodgson and the Reverend John Stevenson, another missionary working among the tribes in western and central India, were the major proponents of common origin theory of all aboriginal languages in India with strong Dravidian influence.[54] Max Müller disapproved of the marginalization of Indo-Aryanism inherent in this idea of Dravidian aboriginality and proposed "the two-race theory of Indian civilization." He suggested an essential racial divide in India, one comprising the Hamites or the Cushites, the other the Japhetites or the Caucasians.[55] The first constituted the tribal population, while the second became the upper-caste Hindus. This two-race theory became the general racial narrative of India by the end of the nineteenth century. The Indian geologist P. N. Bose commented in his note on the tribes of Chhattisgarh, "The people, as elsewhere in India, may be broadly divided into Aryan and non-Aryan, or, perhaps less logically into Hindu and Aboriginal."[56] Verrier Elwin, the twentieth-century missionary and anthropologist who served among the Indian tribes, formally presented in his 1943 book *The Indian Aboriginals* the common theme of Indian aboriginality: contact with Hindus from the plains alienated the tribes from their land and drove them into the remote hills, leading them to poverty and other forms of social and economic deprivation.[57]

The idea of prehistory provided the foundation to this settler-aboriginal racial theme, drawn predominantly from linguistic and ethnological studies. The discoveries of prehistoric burial grounds and humans remains in different parts of central and southern India around the same time as these philological studies provided the deep historical settings to ideas of Dravidian aboriginality. H. W. Voysey was an officer of the EIC who joined William Lambton's trigonometrical survey of India as a geologist in 1818 and surveyed various parts of southern and central India. He was the first among the British to find ancient tombs along the River Godavari in Mangapet in the 1820s, which local villagers believed were built by the *rakshasa*, or demons. He found almost 1,000 slabs of stone, 20 feet square, fixed on upright stones. At their center he found buried sarcophagi containing bones. He speculated that "the antiquity of these remains is very great, beyond all tradition."[58] However, he could not continue his research because of other official duties and ill health. More than 50 years later, William King, the deputy superintendent of the Geological Survey of India, visited the same burial grounds. King seemed unaware that Voysey had found them before, as he commented when he found that the tombs had been opened, "It is possible that this spot has been visited and described by some previous observer."[59] He heard local stories of the site having been "the village of the demons." These demons, he was told, were as tall as trees, unclothed, had long hair, and belonged to a "time beyond the ken of man."[60]

In 1847, soon after Hislop reached Nagpur to start the Central India Christian Mission, he found during one of his missionary tours near the village of Takalghat, around 20 miles from Nagpur, ancient mounds arranged in a circle.[61] He returned to the site a couple of years later and undertook extensive excavations that exposed 90 such circles spread over an area of four square miles. In the center of the circles, Hislop found an iron vessel "like a frying-pan."[62] The ground was covered with little pieces of earthenware, which he speculated were used to preserve human ashes. He also found fragments of pottery and flint. A few years later, in 1854, Hislop undertook another extensive excavation of these sites. The team under Hislop found iron pottery and human remains, and he concluded that these were burial grounds, presumably belonging to a large population.[63]

Philip Meadows Taylor was an army officer for the nizam of Hyderabad. However, he spent most of his time working on geology, archaeology, and engineering projects in various parts of southern and central India. In 1850, he found similar ancient structures in Shorapur, north of the Krishna River, which he described as cairns, cromlechs, and kistvaens. He too inquired among local villagers, who told him that these were built by a "dwarf race" that possessed "great strength" and inhabited the country "in very remote ages."[64]

Meadows Taylor opened a number of these closed cists and found a particular greyish white earth, mixed with portions of human bones, and small pieces of charcoal, along with red and black broken pottery. He also found cairns, which to him were most remarkable, as their construction and the remains inside them allowed "complete identification" with similar monuments in Europe and Central Asia.[65] At Jewurgi (Joldhadgi), in the same Shorapur district, two miles from the River Bhima, he found more cairns. Inside were white-colored earth and human bones; in some there were red and black pottery, a few pieces of decayed spear, and an iron tripod. He noticed that all the skeletons were of short stature, but the bones were of unusual strength and thickness. He found it "curious" that the tradition of the constructors of these monuments being dwarfs prevailed wherever these remains were found and corresponded with the human remains that were found, "indicating, at least, people of low stature." This confirmed to Meadows Taylor the stories of the prehistoric "dwarf race"[66] (fig. 4.1).

Inside the monuments he found human skulls placed upright on a ledge. They were in perfect condition and did not seem to belong to any of the bodies placed in the cist. Meadows Taylor was convinced that they were the "unmistakable traces, and proofs indeed, of human sacrifice." Inside the cist, he found more evidence of human sacrifice: a perfect skeleton with the skull placed at the center of the body. "It had really so ghastly an aspect, that it took some persuasion to get the workmen into the grave to go on with the work."[67]

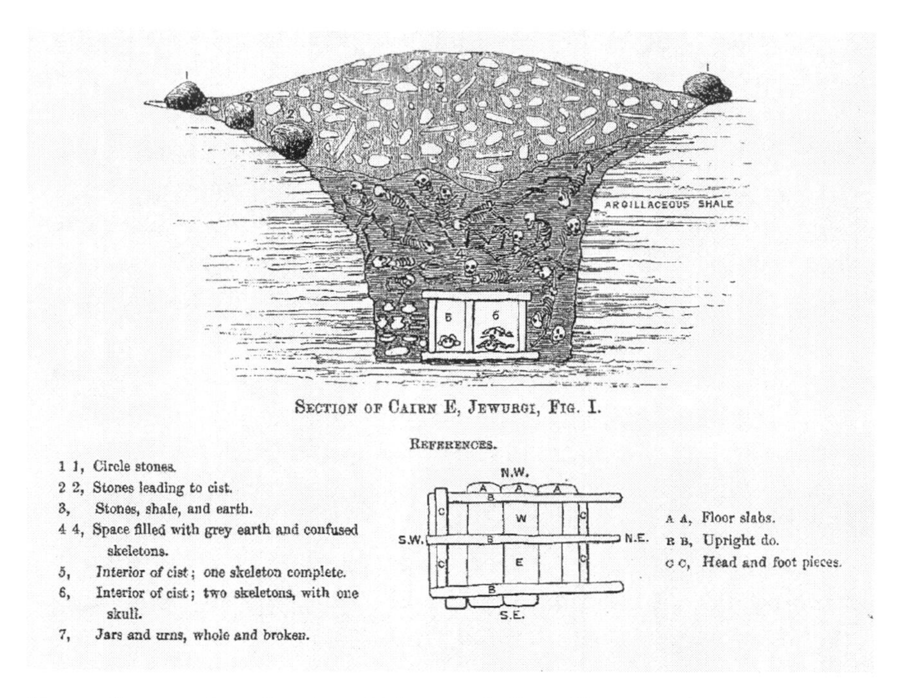

Fig. 4.1. Open section of a cairn with human skulls and bones. Philip Meadows Taylor, *Descriptions of Cairns, Cromlechs, Kistvaens, and other Celtic, Druidical or Scythian Monuments in the Dekhan* (Dublin, 1865), 343.

He produced a sketch of the skull (fig. 4.2). He also found what he described as a "Druidical temple," with rocks piled on top of each other to form columns. He explored several other sites near Hyderabad, where he found similar structures, suggesting strong links with the druidical structures discovered in Europe.[68] The question was, who built them?

As these prehistoric archaeological remains were discovered in central India and theories of Aryan invasion and Dravidian aboriginality dominated the intellectual landscape, ethnologists and geologists were convinced that ancestors of the local tribal populations had built these structures in remote antiquity. These deliberations in India were taking place at the time of and with reference to Europe's own search for its racial ancestry. In Britain, in the middle of the nineteenth century, ideas of Celtic racial heritage, the Celtic fringe, and national Celticism were developing within ideas of an Indo-European common heritage and a strong sense of the universalism of prehistoric humanity.[69] Thus, scholars found links between the colonial Other (the marginalized tribes) and Europe's own Celtic Other. Meadows Taylor's description of these

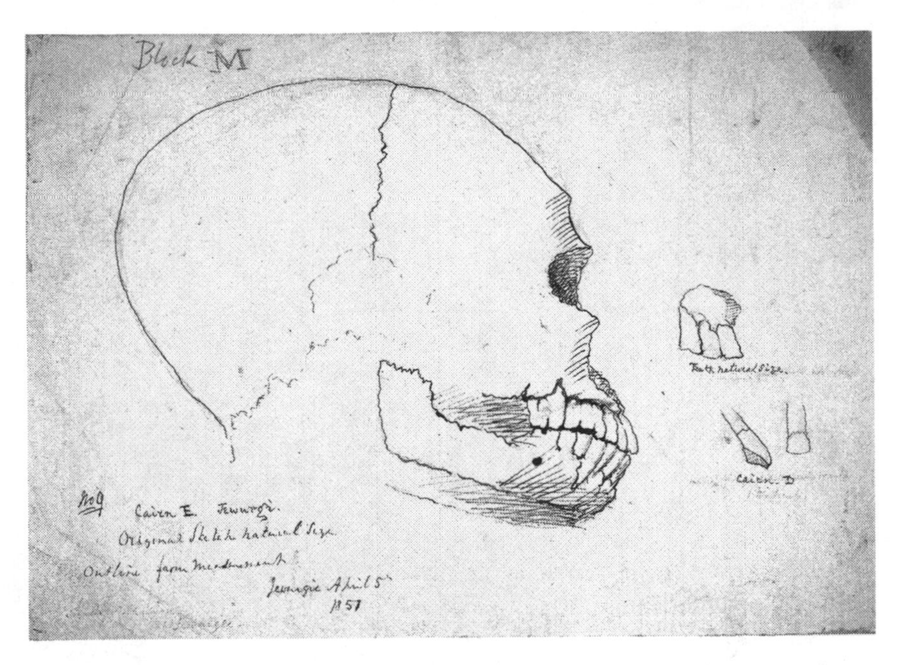

Fig. 4.2. Meadows Taylor's sketch of the "Jewurgi skull." © British Library Board. WD1327 Full-size sketch of a skull from cairn circle grave, Jivarji, Gulbarga District, April 5, 1851.

Indian remains as the "Celtic" and "Druidical" structures of India drew from contemporary discussions in Britain about its Celtic racial heritage. His references to Celtic terms such as "cromlechs," "kistavaens," and "cairns" to describe these Indian structures, and even the way the images were sketched in his book, bore striking similarities to those of Thomas Wright's *The Celt, the Roman, and the Saxon*, published around the same time (1851).[70] Wright suggested that the Celts, who had built the stone structures in Europe, had been driven by the Roman invasion to the remote highlands. Meadows Taylor, in his own paper, which was published in the *Transactions of the Royal Irish Academy* in 1873, suggested that these structures had similar Celtic heritage by making a connection between the Celts and the Scythians, whom Caldwell believed to be the ancestors of the Dravidians. Meadows Taylor believed these races had arrived in India and built similar structures. He quoted from Herodotus about Scythian funeral ceremonies and human sacrifice, and concluded that these graves carried traces of their "deeds of violence."[71]

In 1879, Rivett-Carnac synthesized the work of Meadows Taylor, C. L. R. Glasfurd, Godfrey Pearse, J. J. Carey, Hislop, and Richard Temple on burial mounds and observed the "extraordinary resemblance between the Prehistoric Remains of India and of Europe."[72] He had himself found iron implements, such as bangles, knives, spearheads, and battle-axes that strongly resembled those found with Irish remains.[73] In these discussions, the non-Hindu identities of these races became prominent. Meadows Taylor too assigned non-Hindu and non-idolatry characteristics to these sites. He argued that in some of the cairns in the Nilghiri Mountains in south-central India, Congreve found some idols that Meadows Taylor believed were of later provenance. According to him, these early druidical practices had no idols but had "degenerated" and were "corrupted by Buddhism or Jainism."[74]

This was a return to druidism, referred to in the second chapter, but in a distinct form. While earlier scholars saw eastern druidism as part of the Vedic and Brahmanical traditions of India, later ethnologists, aided by prehistoric archaeological excavations and tribal ethnology, identified druidism with pre- or non-Brahmanical aboriginal histories of India. Wright too stressed that the druids in Europe were part of the Celtic groups.[75] This form of tribal or aboriginal druidism also allowed these scholars to draw parallels between the emergent ideas of Celtic aboriginality in Europe and tribal antiquity in India.

There was one significant difference between these discussions in Europe and India. While scholars in Europe depended predominantly on textual and archaeological traditions in constructing its Celtic heritage, ethnologists in India referred regularly to their study of contemporary tribal populations. This meant that such deliberations on human prehistory in India often gained from anthropological insights. This eventually rendered these connections between India and Europe superficial. While in Europe, as Chris Manias has shown, the "fracturing" of the idea of common origins was caused by the rise of nationalism,[76] in India and Africa it was more due to the emergence of anthropological fieldwork. As more such prehistoric sites were found and compared with the concurrent ethnological observations of central and southern Indian tribes, scholars ascribed these tombs and their builders anthropological lineages. King, who conducted extensive ethnological work among the tribal populations of the Deccan, suggested that the tombs were built by the pre-Aryan Kolarian races that inhabited southern parts of India before the Aryan invasion.[77] He rejected suggestions that the builders of the tombs were of Indo-Scythian heritage. Instead, he pointed to the local Kol tribes (believed to be part of the Kolarian group of tribes) scattered around the region. A potential obstacle to this theory was that the Kol people who lived in the area appeared to have no knowledge of these prehistoric tombs. King explained this lack of awareness as a regression to primitivism: "It is at the same time very remarkable that in the part of the country where I would have it that the evidences of the highest phase of civilization of the pre-Aryan exists, we have now only a very degraded remnant of the race with no knowledge of the ruins in question." Instead, he suggested that the Kolars, the pre-Aryan races of this region who were subsequently driven away by the Aryan invasion, had built the tombs. The contemporary Kols were the less refined remnants of those who returned, while the more refined population were assimilated into the Aryan mainstream.[78] The idea of prehistory as the Other of history corresponded perfectly with the idea of aboriginal tribes living in isolated hills and forests, disconnected from modernity.

Ball similarly observed that the Ho and the Munda tribes, who seemed to be "shut out from all Aryan influences," continued to build similar stone tablets and slabs over their graves in memory of the deceased.[79] Increasingly, from such a strong anthropological orientation, the connections with the distant Celtic race were replaced by connections with the living tribal populations. Edward Tuite Dalton, the chief commissioner of Chota Nagpur and an early anthropologist of the Kol and Munda tribes of the region, published a note on the burial remains there. He described both the "ancient and modern" monuments, as they seemed to belong to the same tradition.[80] His study of the prehistoric stones was an organic part of his anthropological investigations. While examining the stones in Singhbhum District, he found a massive stone slab that had been brought into the neighboring village in anticipation of the death of an old woman. Farther into the forests, he found tribal populations he believed had never seen a white person before. Dalton observed their dances in the evening and in the morning; they helped him clear the forest in search of further burial stones, the prehistoric ones. Deeper in the dense forests, he came across the "very primitive type" of Kols, who according to their own genealogies were the "true autochthons of the country."[81] There he found several ancient burial grounds.

In such oscillations between prehistory and the present, the lines between the two became imperceptible. Thomas Frazer Peppé, a colonial opium officer based in the region, took Dalton to an ancient burial ground known locally as the "place of mourning." He showed him a photograph in which a man, probably a Munda, was made to pose in a mourning posture at a prehistoric burial ground with a backdrop of the hills.[82] In the photograph, the man is shown to be sitting atop a prehistoric stone slab, forming the imagined link between the deep past of human prehistory and tribal life of nineteenth-century India, perhaps gazing pensively at the travesty of time (fig. 4.3).

In his response to Dalton's paper, which was presented at a general meeting of the Asiatic Society at Calcutta in 1873, Rivett-Carnac observed that the former had found a tribe "living in an unfrequented hill-country, which appeared to have practiced from time immemorial, and still to continue to practice, a system of erecting monuments over their dead, similar to the prehistoric remains observed in the hill-country, and comparatively inaccessible tracts of other parts of India."[83] Continuing this playfulness with time and space, others at the meeting pointed out in their comments on these prehistoric sites that the Mundas described by Dalton had distinct "Mongolian" features, which suggested the common origin of the aboriginal populations of

Fig. 4.3. A tribal man sitting on prehistoric burial stones. Edward Tuite Dalton, "Rude Stone Monuments in Chutiá Nagpúr and Other Places," *Journal of the Asiatic Society of Bengal* 42, no. 1 (1873), after p. 118. Image copyright of the University of Manchester.

Chota Nagpur and those of the Burmese-Malayan region. They argued that several words in the Kolarian languages were similar to those of the "Talaings" (Mong) of Pegu and of the people of Mong La region in Cambodia.[84]

As the anthropological frame became the dominant explanation of prehistoric remains, explorers in India found connections with tribal populations in other parts of the world. Richard Temple, who was the chief commissioner for the Central Provinces between 1862 and 1867, wrote that the Gonds had much in common with similar "wild races" of other parts of India and Asia.[85] Dalton confirmed these suggestions in his work on the Mundas.[86] Others observed that the inhabitants of these hills had remarkable affinities with the South Sea Islanders: "They are ignorant and superstitious of course—but they are without the bondage of caste and priesthood. . . . They are simple but independent & fearless people—and very trustfull. Men of their word, who may be perfectly depended on—tho' they are greatly given to the intoxicating drink."[87] German missionaries, such as Bernhard Schmid, who worked in India suggested that African Hamite tribes came to India and populated the South Pacific Islands as well. The missionary W. Taylor agreed with Schmid on similarities between the southern Indians and the South Pacific Islanders (particularly the Samoans and Tongans), among whom his fellow missionaries had worked as well. Taylor suggested that "southern races" spread to the Northern Hemisphere prior to any migration of northern people toward the south.[88] These ethnolinguistic and archaeological studies gradually contributed to theories of Austric and proto-Australoid races that eventually traced the history of a "southern" tribal migration and a "southern" trajectory of aboriginality and human prehistory that preceded Aryan migration.

Theories of a southern migration of races originated from linguistic and ethnological studies conducted mainly by a network of missionaries and ethnologists in central and northeastern India, the Strait of Malacca, and the Pacific Islands. In India, this was a complex and occasionally inconsistent theory that had both biblical and Dravidian lineage but was essentially posited against its Aryan traditions. These became part of a larger theory about the racial and historical similarities among aboriginal populations in India, Southeast Asia, and the Pacific. In 1870, T. H. Huxley suggested that the "hill-tribes" of the Deccan were similar to the aboriginal Australoid races of Australia:

> The only people out of Australia who present the chief characteristics of the Australians in a well-marked form are the so-called hill-tribes who inhabit the interior of the Dekhan, in Hindostan. An ordinary Coolie, such as may be seen

among the crew of any recently returned East-Indiaman, if he were stripped to the skin, would pass muster very well for an Australian, though he is ordinarily less coarse in skull and jaw.[89]

Around the same time, the American anthropologist Lewis Henry Morgan suggested that the supposed similarities between Iroquois and Dravidian kinship reflected similarities across Asia and Oceania, thereby providing a common ethnic heritage between Asian and Polynesian aboriginal races and Amerindians.[90] Deriving from these suggestions, the American anthropologist Roland Burrage Dixon first proposed the term "proto-Australoid" in 1923 in his book *Racial History of Man*. It was supposed to be an ancient race of humans, primarily hunter-gatherers, who descended from the first major wave of humans to leave Africa.[91]

In India, in the 1930s, the idea of the proto-Australoid acquired significant political and historical value as it linked assertions of tribal rights and identity with those of the deep history of humanity in the subcontinent. Prominent Indian anthropologists such as Biraja Sankar Guha (founding director of the Anthropological Survey of India, 1945), studied the recently discovered human skulls of the Indus Valley civilization and suggested that they belonged to the proto-Australoid groups. He then connected them with Meadows Taylor's sketch of the "Jewurgi skull" (which he thought was "definitely negroid"); the skulls of the Mundas of central India, the Veddas of Sri Lanka, and those of the Aborigines of Australia; and various megalithic monuments as evidence of the extinct and extant traces of proto-Australoid races and cultures.[92] Others traced several hill tribes in central and southern India as living specimens of proto-Australoid races, suggesting a racial continuity in India from the Indus Valley civilization to the contemporary tribes.[93] This link with the Indus Valley people became part of the political assertion of Adivasi identity in postcolonial India. Jaipal Singh Munda, the tribal leader of the Jharkhand movement, announced in his famous speech to the first Constituent Assembly of India on the eve of Indian independence, in December 1946, that "newcomers" had "driven away my people from the Indus Valley to the jungle fastness. The whole history of my people is one of continuous exploitation and dispossession by the non-aboriginals of India."[94]

Michael Witzel's account of the emergence of Gondwana mythologies, covering the southern aboriginal cultures of Australia, the Andaman Islands, Melanesia, and sub-Saharan Africa, is loosely based on these nineteenth-century ethnological and linguistic theories of the southern races.[95] Shaped by the imperial globalism of late nineteenth and early twentieth centuries, these theories suggested a history of aboriginality through prehistoric migrations of humans across the global south. The discovery of human remains in the Narmada valley in central India in 1982, commonly referred to as the "Narmada Man," rekindled the idea of supposed links between prehistoric humans and tribes.[96] These human remains have been described as belonging to "short and stocky" hominins who may have migrated to Asia from Africa. Linking once again prehistory with anthropology, scientists conducted DNA tests among the Mundas to find that they too carried the same "short and stocky" traits of these hominins along with "Andaman pygmies," which suggested the continuity of these "short-bodied" populations in South Asia.[97] This contributes to the modern "southern race" theory, which is now an established scientific idea of the earliest human migration, supported by DNA research, suggesting a coastal exodus of the earliest humans via the Indian Ocean rim around 60,000 years ago.[98]

The two-race scheme of the aboriginal and the Hindu/Aryan invader or settler placed the former as also the living specimens of prehistoric humans. In this global scheme, tribal populations were situated against the settlers or the migrants, and human prehistory was equated with aboriginality. The assumption was that these people had lived on the earth from the dawn of civilization almost unchanged. This idea appeared more credible, as it seemed to corroborate the laws of nature, where the landscape appeared equally prehistoric and where certain contemporary species resembled fossilized ones. The catalog of the London Exhibition of 1873 explained, "As among existing species we find the representatives of successive stages of geological species, so amongst the arts of existing savages we find forms which, being adapted to a low condition of culture, have survived from the earliest times."[99]

As the links between aboriginality and prehistory were being established from the middle to late nineteenth century, tribes were often exhibited as living specimens of early humans.[100] One of the earliest ethnological exhibitions was held in Jabalpur in central India in 1866. The organizers displayed what was described as "curious" specimens of the human race. While planning the exhibition, the chairman of the Asiatic Society of Bengal suggested to Temple, the chief commissioner of the Central Provinces, that since the exhibition was being held in a part of India with large tribal populations, the display should include a live "human department." Temple passed on the suggestion to the organizer, Colonel T. Spence, and asked whether they could also be shown more "authentically" as savages or primitive humans, "to see whether a little lucre may not tempt these wild creatures to come into the station and be clothed, and shewn off for the edification of their more civilized fellow-humans."[101] Spence, who acted as an intermediary between the tribes and the colonial authorities in the region, was not very enthusiastic, as he feared that "they are as wild as the jungles and the hills which they inhabit" and could not be tamed for such a display. He recounted that the "specimens" of the Baiga tribe "boasted nothing more in the way of clothing, than a green tassel, and a powder-horn, which, however, cool and airy, was scarcely sufficient for decency!" He promised to try to get as many people from different races as possible and "should any scientific men desire to make an examination of the heads and general conformation of any of these specimens of the human family, our Committee will give all the assistance that can be rendered without risk of causing annoyance or apprehension."[102] The chairman of the Asiatic Society insisted that they ought to be displayed in their "natural" attire for the sake of authenticity: "As cleanliness comes after godliness, so I think that decency must come after science."[103] Anthropology had established their state of nature, even if that was not always decent.

The Jabalpur exhibition was followed in 1874 by the anthropological exhibition at Bethnal Green organized by the South Kensington Museum, based on the collections of the English army officer and ethnologist Augustus Henry Lane Fox (better known as Augustus Pitt-Rivers). British ethnologists sent clay models of Indian aboriginal tribes to the exhibition.[104] The catalog, which guided visitors through the displays of human skulls, models, hair specimens of modern "barbarous races," and then to the paleontological collections of remains of early humans, explained the rationale of finding the prehistoric among the "modern savage":

> The resemblance between the arts of modern savages and those of primeval men may be compared to that existing between recent and extinct species of animals. As we find amongst existing animals and plants species akin to what geology teaches us were primitive species, and as among existing species we find the representatives of successive stages of geological species, so amongst the arts of existing savages we find forms which, being adapted to a low condition of culture, have survived from the earliest times, and also the representatives of many successive stages through which development has taken place in times past.[105]

Two things are apparent here. First, the use of the term "modern savage." Lubbock used it originally to refer to the "primitive" state of aboriginal populations in South Africa and the Pacific Islands.[106] It was now a normalized category to describe tribal populations as direct descendants of prehistoric humans, clarifying the connections between primitivism, aboriginality, and prehistory, all invested in living tribal populations. Second, the use of the phrase "primitive species" to refer to both nonhumans and humans, living and extinct. Linnaean classification had rendered humans, particularly "savages," along with nonhumans, as biological species.[107] The concurrent use of the phrase "geological species" here connected that biological definition with the prehistoric one. In short, deep history viewed tribal and marginalized populations as simultaneously biological and paleontological entities.

Humans as Apes

The designation of tribal populations as both biological and geological species resulted from a particular intellectual intersection in the nineteenth century. Alongside the aboriginal-prehistoric trope that was common in nineteenth-century paleontological and anthropological discussions was another convergence between zoological and anthropological studies of primates and humans, respectively. There were two different trajectories in this deduction. First, there were the joint explorations by geologists and anthropologists for human origins in which they studied tribes and the paleontological remains of humans simultaneously for clues of human evolution. The other was in the search for biological evolution of humans from primates, shaped predominantly by Darwinian ideas of evolution. In his *Descent of Man* (1871), Darwin claimed that humans were descendants both of apes and of "barbarians."[108] For the latter, he referred to his ethnological observations of Fuegians, mentioned at the start of the chapter.

While Darwin established the formal links between the evolution of humans and primates, primarily in *Descent of Man*, these debates were already taking place in a much wider intellectual forum. The question of primates and their relationship with humans had occupied zoologists, paleontologists, anatomists, and finally anthropologists throughout the nineteenth century. As Darwin himself suggested, from the 1860s, there were deliberations among scientists about the similarities between humans and apes.[109] T. H. Huxley, in his *Evidence as to Man's Place in Nature* (1863), demonstrated that humans were not unique as a species and had indeed evolved from primates. While Huxley's point was based mainly on zoological examinations of primate anatomy, Charles Lyell in his *Antiquity of Man*, published in the same year, made a similar point, that humans emerged from apes, based on paleontological observations of fossilized skulls of early humans. Even before Lyell, in the 1850s,

Falconer revisited his earlier discoveries of Siwalik primate fossils in India to assert that they were the earliest discoveries of human prehistory. He reassessed his finding of a primate jaw and teeth in the Siwaliks in 1837 and concluded that the primate fossils of the Himalayas held secrets of early humanity in the Indian subcontinent.[110] He also noted the resemblance between the primate fossil tooth in his collection and the tooth of an orangutan in the collection of the Asiatic Society of Calcutta as well as the living monkey populations in the hills. He then connected these to his discussions on the "primeval man" to suggest that in India, "every condition was suited to the requirements of man."[111]

There was a particular episode when these zoological and paleontological inquiries came together. In 1856–57, German paleontologists found a fossilized skull in a valley of the River Düssel near Dusseldorf, which was believed to have belonged to an archaic human species. They named the species Neanderthal, after the valley where it was found. Lyell visited Germany and carried the cast of the skull back to England and showed it to Huxley, who had been working with skulls of primates. As Huxley held the cast of the several-thousand-year-old skull in his hand, the deep past of human paleontology collapsed into the primate zoology of the present. He immediately confirmed to Lyell that the Neanderthal skull was the "most ape-like skull he had ever beheld."[112]

As a consequence of these paleontological discussions of human antiquity and the growing zoological knowledge of primates by the 1860s, several human populations were compared to apes. European scholars freely compared the brains of primates with those of aboriginal populations, such as the Bushmen of southern Africa.[113] The same skull and skeleton discovered at the Neandertal valley were exhibited at a German scientific meeting in Bonn in 1857. Scientists, while certain that they had very little similarity to European skulls and skeletons, debated their resemblances to those of indigenous populations living in South America.[114] Huxley himself found "marked resemblances" between these fossil skulls and those of the Australian Aboriginal population. To stress his point he even suggested that the stone ax was "as much the weapon and the implement of the modern as of the ancient savage."[115] By the 1890s, it became common to compare Australian Aborigines with animals, mainly drawing from Darwinian evolution.[116] On the one hand, zoologists found primates (fossilized or living) approaching humans. On the other, paleontologists and anthropologists found humans (again, both fossilized and living) who appeared apelike.[117]

In the mid-twentieth century, the sociologist Marshall Sahlins compared "societies" of monkeys and apes with those of aboriginal people in Australia,

Polynesia, Congo, Greenland, and other parts of the world to test the assumption that "these [simian] societies parallel early cultural society in general features."[118] Sahlins's project was different from that of the evolutionary scientists and anthropologists of the nineteenth century. He sought to identify the origin and prehistory of human societies, not that of the human species. In his conclusions, Sahlins privileged culture over biology in shaping human behavior, stressing, "Human social life is culturally, not biologically, determined."[119] His studies also established several discontinuities between primate and aboriginal societies. Yet, even this cultural frame, premised on human "agency" and his conclusion, "the emergence of human society required some suppression, rather than a direct expression, of man's primate nature," was founded on the comparison between aboriginal populations, prehistoric humans, and primates.[120] The very conception of Sahlins's project, more specifically his search for "man's primate nature" in aboriginal populations, highlights the problem I am exploring here: the prospect of a continuum between the aboriginal, the prehistoric, and the primate.

Both these frames were applied to tribes in India in the nineteenth century, suggesting that they had close affinities to *both* prehistoric humans and primates. At the same time, in India, these developments took place both within and without the Darwinian frame. Since the early colonial encounters with tribes were mediated through the Hindu textual references to them as demons or the lowest forms of humans, the prehistory of tribes and their connections with primates evolved primarily from references to Hindu textual and caste traditions. These references to humans as primates were derived from the dehumanization of marginalized populations inherent to Hindu caste traditions.

In 1820, Voysey crossed into central India while conducting his surveys and climbed up the "Sugriva Perwattam" (the hill of Sugriva). Sitting on its peak, according to legend, Hanuman, the god of monkeys, recruited his army of monkeys for Rama. Voysey alluded to characters from the mythological epic *Ramayana*, which narrates the tale of Rama, the exiled king from the north traveling into southern and central India. The story narrates the abduction of Rama's wife Sita by demons. In retaliation, and to rescue Sita, Rama formed an army of monkeys with the help of Hanuman in these dense forests and hills of central India. As Voysey himself sat atop the hill, surrounded by monkeys, he sardonically noted, "Certainly he [Hanuman] chose an excellent place, the monkies being very abundant."[121]

While Voysey found himself surrounded by simian populations in the hills of the mythological monkey kingdom of the *Ramayana*, others found humans who resembled these mythological apes. In 1824–25, Henry Piddington, merchant, coffee planter, meteorologist, and curator of the Asiatic Society Museum in Calcutta was establishing a coffee plantation in the Palamau District in western Bengal at the eastern end of these central Indian hills and forests. He gathered laborers from local Dhangur and Kol tribes. One day, he found that several villagers had flocked to see a small group of laborers among them, whom they called the "monkey people." Intrigued, Piddington sent for them and found that they "justified the epithet which the villagers had applied to them." He described a short, flat-nosed man with semicircular wrinkles around the corner of his mouth and cheeks and disproportionately long arms. He thought that in the dark, this man could be mistaken for a large ape: "In the gloom of a forest, the individual I saw might as well pass for an Orang-Utang as a man." He wanted to send him to Calcutta for Clark Abel (founder of the Calcutta Phrenological Society) and the Asiatic Society to inspect. However, the man in question became suspicious and ran away in the middle of the night, never to be seen again.[122]

Left to his own means, Piddington resorted to his reading of the *Ramayana* and compared the mythology with his field experience. His small coffee plantation provided an allegorical setting for the myth of the monkeys. He deduced that there was a small forest tribe, somewhere in the "wild country" between Palamau, Sambalpur in the east, and the source of the Narmada River in the west, who were remnants of the prehistoric races of India. He believed that the *Ramayana* allegorically referred to these tribes as the monkeys who formed the army of Rama.[123] Piddington subsequently, in 1855, raised this with Falconer, who was then in Calcutta. Falconer informed him that in London he had met George William Traill, who had heard similar accounts of a race of people in the dense forests of the Terai in the Himalayan foothills who resembled monkeys and lived in the trees. Traill once managed to get hold of one of them and found they justified the phrase *bon-manush* (literally "forest people" but also apes) the locals used for them. Piddington ended his note by asking, "What are these singular people?"[124]

Both episodes around the human-ape continuum took place in the 1820s, long before the evolutionary deliberations started in Europe. In India, discussions of the human-nonhuman continuum followed two lines. Since the eighteenth century, Orientalists had observed depictions of human-nonhumans in primarily Hindu religious iconography, deities, and mythology. They noticed that in the *Ramayana*, Rama, the protagonist, formed an army of monkeys to invade the kingdom of Ravana. Based on the allegorical reading of

Ramayana, they assumed that the monkeys represented the aboriginal races of India. In such a reading, they were aided by the Hindu texts that treated tribal races as the lowest forms of humanity in the broad spectrum of the Hindu caste system. Ethnologists, therefore, made the connection between Indian tribes and apes, as referred to in mythology.[125] By the 1870s, the mythological monkey and aboriginal human schema was well entrenched in India. In this ethnological reading of the *Ramayana*, the Aryans, who were believed to have composed the *Ramayana*, referred to aboriginal tribes of India as monkeys because they identified their physical features and their inhabitation in dense forests with those of the simian populations.[126] Inherent in this is the dehumanization of the tribal population within the Hindu intellectual and social corpus.

The second development took place when this allegorical reading was fused with the Darwinian framework of human evolution from apes. In the early twentieth century, ethnologists such as Edgar Thurston drew from both Hindu texts and ethnological studies in India to affirm much clearer arguments that the presumably primitive races of India had a close affinity with apes, from whom they had supposedly evolved.[127] In the process, Brahmanical characterization of the tribes as subhuman races came to be referred to in Darwinian terms.

Piddington was straddling yet another aspect of the British encounter with the ape-human theme in nineteenth-century India with his use of the term *bon-manush*, which carried the dual meanings of "forest people" and "apes." Along with the mythological allegories, the British also noticed "strange" proximities between humans and animals in everyday lives in India, where keepers of animals often lived with their livestock or humans revered and worshipped animals.[128] They also encountered the "wolf-boys," or children who were supposedly nurtured by wolves or other animals in the wild, of which Rudyard Kipling's Mowgli in *The Jungle Book* was a poignant representation. Therefore, depending on the context, *bon-manush* could mean an ape, an ape-like prehistoric human, or humans who have grown up in close proximity to apes or wild animals.

This layered implication was evident at the Jabalpur exhibition of ethnology held in 1866, where the term *bon-manush* was used with much clearer zoological connotations. They were classified under the "birds and beasts of Hindustan" and with a clear description: "The Bunmanus is an animal of the monkey kind. His face has a near resemblance to the human; he has no tail and he walks erect. The skin of his body is black, and slightly covered with hair one of these animals was brought to His Majesty from Bengal. His actions were very astonishing."[129] Yet the "Bunmanus," so designated and exhibited there, did not belong to the so-called wild or aboriginal tribes but were men who had been nurtured by wild beasts.[130] Therefore, human-animals could represent, depending on the context, diverse forms of imagined aberrations of humanity.

It is true that in India the human-primate schema did not necessarily develop from Darwinian ideas. However, its social implications are no less problematic than the search for the so-called missing link among certain human populations.[131] The ape-human continuum in India was constitutive of the dehumanization of marginalized groups within the Indian caste system and agrarian economy, the thematic implications of which were adopted by Europeans. The subordinated Dalits (a term used in twentieth-century India to refer to members of the scheduled castes, or untouchable communities) of India were often known by the term *pariah* (literally "outcast, despised, and untouchable"), which Rupa Viswanath has accurately described as "cruel," for the revilement of the people that it signifies.[132] The same term was used by Europeans to describe the stray dogs of India, denoting, according to the *Hobson-Jobson*, "a low-bred casteless animal."[133] Therefore, the human-nonhuman, or more specifically, the ape-human continuum in India was also steeped in the dehumanization inherent in the caste system. While references to ape-humans in India in the nineteenth century did not always have evolutionary connotations, they were no less derogatory and were usually reserved for tribal populations. This dehumanized status of the tribes was subsumed within Darwinian evolutionary ideas. In the process, Indian tribes were seen as representations of dehumanized marginalities that belonged to both prehistory and evolution.

Tools of Prehistory

The dehumanization inherent in the zoological connotations of the tribes explains other aspects of the tribe-prehistory continuum that Europeans imposed on the marginalized and subordinate populations of India. British ethnologists, geologists, and explorers saw in these tribes, who they believed had been driven to the hills and forests by more civilized races, a form of material life that was simple, primal, but also marked by deprivation. They imagined the same deprivation to be true for life in prehistoric times. In other words, tribes in the isolated hills and forests, they assumed, had continued with their destitute forms of existence, away from agrarian and mineral resources, since prehistory.

Ball described a particular incident that illustrates the nineteenth-century imagination of human prehistory as a form of de-humanity. During his travels in central India along the Narmada valley in search of coal deposits, he found some chipped stone implements in the dirt that had been dug up. He surmised that these were prehistoric tools used to dig wild roots from the ground for food. These tools reminded him of a woman "of the lowest type" whom he had seen doing so with a similar tool. His description of this rudimentary act in search of food created a primal link between the deprivation of this wretched woman and prehistoric existence: "As I saw her, with hunger in her eyes and an infant strapped on her back, while she crouched over the precious root which she was digging out, I could not but regard her as being in all probability a lineal descendant of the manufacturers and users of stone implements similar to those which are figured."[134]

Ball's paleontological investigations of prehistoric stone tools were invariably accompanied by similar ethnological observations of "aboriginal races." One morning, a large bear attacked his camp, which he promptly killed with his rifle, while, as he described, still smoking his cigar. In the evening, as he returned from work, he supervised the skinning of the animal. The tribal people had been given knives for the job. However, he noticed that they soon discarded them in favor of their small axes, which they removed from their handles and used "with much greater efficiency than they could the knives." Ball could not help thinking that "the axes so employed represented the stone skin-scrapers of pre-historic times." He compared these tools with the prehistoric stone implements that he had found in the neighboring Satpura hills, which, he now became convinced, had been used "as skin scrapers" in prehistoric times.[135] He concluded that "the various forms of traps and snares which are now commonly met with in the jungles may be survivals of the ancient methods which were employed to capture the wild animals."[136]

The discovery of prehistoric flint tools occupies a central place in the exploration of human prehistory. These, along with human fossils, provided geologists with clear material evidence to date and trace the evolution of early humans. The tools, as they were found in different strata, were also one of the key objects through which humans were situated within the deep history of the landscape. Boucher de Perthes of Abbéville first suggested a link between flint stones and human history in the 1840s. He collected several tools found during the dredging of the Somme River canal and proposed that they proved that human history preceded biblical narratives, as they came from the lowest strata of the alluvial deposits of the valley.[137] By the 1850s, geologists and archaeologists were also beginning to accept the contemporaneity of humans with extinct fauna more generally. Falconer left for Europe from India in 1855. In 1858, he saw the flints found by Boucher de Perthes and became convinced that they were designed by ancient humans who lived alongside mammoths. He soon started his own research in the Brixham Cave in Devon and urged fellow geologist Prestwich and the archaeologist John Evans to investigate further in the Somme valley. Lyell too traveled to Abbéville in 1859 to see the flint stones and was convinced about the prehistory of humanity. These led to the emerging consensus in Europe around 1859.[138] Reflecting on this wider change in the history of human prehistory through the discovery of stone tools, the French paleontologist Marcellin Boule commented in 1923, "Barely a century and a half have elapsed since the prime question of the origin of Man was raised from the regions of dreams and fiction to the domain of science."[139] For metropolitan scientists such as Boucher de Perthes, Lubbock, and Lyell, the significance of the stratigraphic location of these flints in the landscape was in their contemporaneity with nonhuman species and the challenge that posed to the prebiblical emergence of humans.[140]

Crucially, in South Asia, stone tools also linked paleontology with tribal anthropology. As we have seen in the case of Ball, paleontologists and geologists in India discussed these tools with reference to the tribal populations living in the neighboring regions. More generally, for those in Asia or Africa, *because* of their proximity with tribal populations and the adjacent questions of tribal aboriginality, the stratigraphic location of these tools became entangled with questions of the indigeneity of these people to the land and their current states of existence.

The term "metropolitan" above is used consciously, as even in Europe's own peripheries marginalized populations and their seemingly primitive artifacts were seen as remnants of prehistoric humanity. Linda Andersson Burnett has shown that the Swedish zoologist and archaeologist Sven Nilsson, in his *Skandinaviska Nordens ur-invånare* (*The Primitive Inhabitants of Scandinavia*), published between 1838 and 1843, argued that a study of the "primitive" archaeology and tools of the indigenous Sami people of Scandinavia could provide insights into the remote human past.[141] In Britain, in contrast, the immediate context for these tools was contemporary industrial culture, from which Evans drew to explain and classify the flint stones.[142] Therefore, the discussions on flint tools in India took place in a different context, with reference to both living aboriginal populations and the geological formation of the landscape where these were found. There prehistoric flint tools were discussed in concurrence to three themes: the tools themselves, the geology of the landscape, and the aboriginality of the tribal populations living nearby.

For example, in 1870, the commissioner and ethnographer of Central Provinces, Charles Grant, in his *Gazetteer of the Central Provinces*, referred to the flint knives and cores found in the Narmada region that suggested the early evolution of humans at these sites.[143] He immediately referred to the prehistoric landscape of the Central Provinces, which he believed held the clue to exploring "this intensely interesting question—the antiquity of man."[144] The seeming agelessness of these forests presented the possibility that humans had lived there since prehistoric times. The conditions in which the prehistoric "*Hippopotami* wallowed in the muds, and the *Rhinoceros* roamed in the swampy forests of the country, where *Mastodons* abounded," had remained relatively unchanged in the modern period. Therefore, Grant asked, "Was not man also contemporary with these now extinct animals?" He replied, "There is not a shadow of proof that the country was not then, as now, fitted for the abode of man."[145]

Grant was most probably referring to the first major find of stone tools in the region, by Downing Swiney, a lieutenant of the Royal Engineers. In 1864, Swiney found flint implements in the gravel of the Narmada River basin near the city of Jabalpur. They appeared to be significant, as Lubbock and Lyell in Britain had around the same time established flint instruments as crucial clues to the question of prehistoric human material lives. Swiney compared some of them with those mentioned in Lyell's *Antiquity of Man*, published just a year previously, and the rest he sent to Lyell in England. Swiney found that although the hammers and knives were similar to those found in Europe, some of the polygonal stones were distinctive. His paleontological inquiry into the age of these stones had an immediate anthropological point of reference. As he speculated about the age and use of these polygonal stones he inquired among the local tribes, who declared that they were just pebbles. Swiney was unable to continue his research, as he died soon after, in 1865. Rivett-Carnac, who was the secretary of the Antiquarian and Scientific Society of the Central Provinces, which had been formed in 1863 in Nagpur, took up the inquiry. He initially speculated that the polygonal stones were remaining "cores" out of which the knives and arrows had been carved, similar to those found in Europe. He subsequently rejected the idea, as they did not match the tools found in the vicinity. He then suggested that humans who were in the most primitive stage and had not yet learned how to smooth and polish their weapons had made them. The fact that they were found in the Pliocene strata of the

Narmada valley indicated that the humans who had originated in these parts of India were far older than those found in Europe or in America.[146]

Swiney, Rivett-Carnac, and others formulated these ideas of deep human prehistory in India with reference to the adjacent ancient rocks, rivers, and tribal populations. These suggestions were often incompatible with metropolitan postulations, which still placed Europe at the center of human evolution following Lyell's propositions in *Antiquity of Man*.[147] Rivett-Carnac sent these polygonal stones to the Edinburgh Geological Society to ascertain the truth of his speculations about their deep history. The society consulted John Evans of the British Museum, who had received Swiney's previous batch of flint stones as well. Evans had traveled to France in 1859 to observe Boucher de Perthes's finds and subsequently played a significant role in Britain in establishing the role of stone tools in human prehistory. Evans was unconvinced of Swiney's and Rivett-Carnac's speculations about the Indian tools. He suggested that they were chipped cores from knives or small arrows and thereby were not any older than the knives.[148] From London and without any clear knowledge of the landscape, he rejected Rivett-Carnac's suggestion that these stones resembled the gravel of the Narmada valley and could, therefore, belong to the Pliocene era.[149] The Edinburgh Geological Society confirmed Evans's views that these "cores" were "precisely similar to the flakes found in Europe"[150] The society even speculated that Swiney had come across an abandoned workshop of implements of the "native tribes at Jubbulpore."[151]

While rejected in Britain as indicators of deep human antiquity, the Jabalpur implements remained important points of reference to human prehistory in India. The following year, in 1866, they and several other implements were discussed at the meeting of the Asiatic Society of Bengal. W. T. Blanford presented the "Jubbolpore tools" to the society. He was aware of the metropolitan verdict that the stones were not as old as suggested by Rivett-Carnac and Swiney but maintained, "I am much disposed to believe that we have evidence in India of the existence of man at a much earlier period than in Europe." He concluded that this conjecture "has not attracted the attention it deserved."[152] In 1867, Blanford left India briefly to join the Abyssinian expedition of the British army and became preoccupied with drawing comparisons between Indian and African fauna. The question of human evolution in central India receded from his plans. However, throughout the nineteenth century, discussions continued in India regarding the early origin of humans in the subcontinent based on the discoveries of flint tools.[153]

Robert Bruce Foote's discovery of Paleolithic artifacts in southern India in 1863 is regarded as a watershed moment in the Indian history of the discovery of human prehistory. Foote, who had served in the Geological Survey of India since 1858, discovered a stone hand-ax, made of quartzite, on the Parade Ground at Pallavaram Cantonment, close to Madras, in May 1863. Foote was inspired by Falconer, Boucher de Perthes, and Prestwich in his search for traces of human prehistory in India. He subsequently found other implements in different parts of Madras Presidency, along with William King.[154] In some of the places, they found kistavaens and stone tombs as well.[155] These discoveries received significant attention in Europe at the time, as they appeared to confirm the finds of similar implements in the Somme valley in France and in England.[156] King, who had accompanied Foote on his excavations, presented in detail their discoveries at the Asiatic Society of Bengal in 1867, at the same meeting that W. T. Blanford discussed Swiney's tools.[157] King had also inquired among the most "uncultivated people" living nearby, called the "Chensulahs," who could not remember whether their ancestors used them.[158]

Scholars who regard Foote's discoveries as a pioneering moment in the discovery of Indian prehistory, tend to see them in isolation, away from the wider questions of aboriginality that shaped these ideas about the prehistoric tools.[159] The key feature of Foote's discovery was the way he situated these tools alongside existing theories of aboriginality and of the geological formation of the Indian subcontinent. In other words, he provided the tools with a natural-historical setting. In their first publication on these stone implements, Foote and King were quite vague about the people who might have made or used them, referring to them as loosely "ancient men," without any clear anthropological insights or chronological context.[160] He admitted that Paleolithic humans did not survive into the present age.[161]

Foote was more forthright in identifying these tools with specific human races and geological evolution in his later publications. He visited Britain in 1868 to attend the International Congress of Prehistoric Archaeology held at Norwich. During his visit he presented his paper at the Geological Society, where he showed with the help of a large map the physical geography of the region where these tools were found (fig. 4.4). In his note to the Geological Society, he placed them and the stone circles and kistavaens within the grand narrative of geological evolution of the subcontinent and the emergent theories of Dravidian aboriginality and human migration in peninsular India. Most of the implements were found along with laterite deposits. He explained

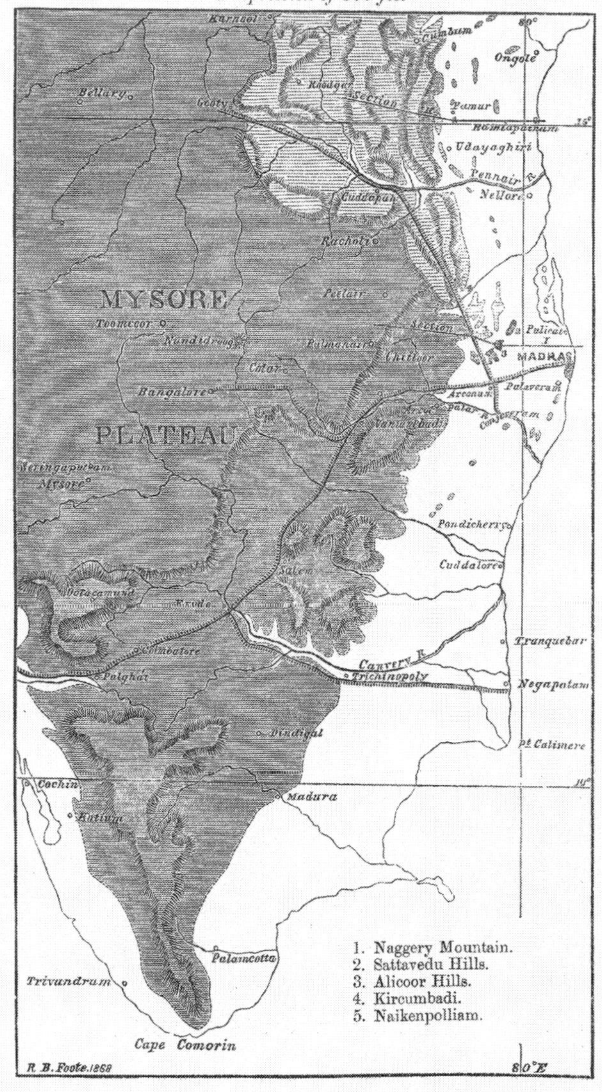

Fig. 4.4. The laterite formation of the Deccan and sites of the stone implements. Robert Bruce Foote, "On the Distribution of Stone Implements in Southern India," *Quarterly Journal of the Geological Society of London* 24 (1868): 485. Image copyright of the University of Manchester.

that there were organic remains found in the laterite formations, which were found in the entire southern and central peninsula of India. He suggested that this entire region was once submerged in water, which led to the laterite formation, which was gradually raised above by movements of the surface.[162]

The stone implements, which were all purple-brown in color from being embedded in laterite, were found in the hills at a height of 500 to 300 feet above sea level. He suggested that these hills, populated by prehistoric humans, were once isolated islands that were later joined to the Indian mainland.[163] Foote's geological narrative of isolated hills as sites of human prehistory conformed to the hill-aboriginality schema of the nineteenth century created by colonial ethnologists, missionaries, and philologists. T. H. Huxley, the president of the Geological Society, agreed in his response to Foote's paper that the region was once an island. He then connected this conclusion with ethnological research elsewhere to suggest that the people who lived in this island were not Aryans but "the ancestors of the Hill tribes, whose nearest affinities are with the aboriginal Australians of the present day."[164] He believed that the two populations were "nearly or quite continuous, having been subsequently cut into segments by geological changes—and that the makers of the quartzite implements came from the same stock as both these recent tribes, which present the most rudimentary civilization known."[165]

During his exploration for these tools, Foote consulted local tribal populations living in these seemingly isolated hills. Once in the hilly region of Kurnool District, he showed them to a man from the Yanadi tribe who had wandered up to their tent and asked him whether the people of his hill tribe were familiar with them. The experience did not go well: "In reply, he gave me a look of the most withering contempt and marched off in a stately way, absolutely refusing to come back and answer any more questions. He was evidently much offended by my question, though I had spoken quite kindly to him, but he got immediately on to a very high horse to my great amusement."[166]

In his later publications, Foote summarized the theoretical deductions around the entire collection of several such implements across India. He divided the tools into Paleolithic and Neolithic ones. Other pottery belonged to early and later Iron Age.[167] He suggested that aboriginal people, whom he called "Deccanites," lived in the Deccan region of central India in the prehistoric period.[168] He noted that there were differences in opinion about the "Dravidian races." While some believed that they were migrants from the northwestern regions, others such as Herbert Hope Riseley suggested that they were the "autochthons of the Deccan."[169] Foote's Dravidians were different from

Caldwell's, which had been based on textual and philological studies. They now became a more geologically defined Paleolithic race, who, according to Foote, entered India from the northwest and passed through the Indus Valley to the Rann of Cutch, and then southward to the Tapti valley, where they entered the Deccan Plateau region.[170] He associated the route with the changing course of the Indus and the drying up of the Rann as well as from the Paleolithic remains that had been discovered. He suggested, conforming to the emerging southern race theory, that they could be the "brown race" described by Elliot Smith in *The Ancient Egyptians* who were presumed to have migrated out of Africa and populated various parts of the Southern Hemisphere.[171] He believed that the study of the craniology of the Deccan people would confirm these suggestions.[172]

Such linking of stone tools with contiguous geological and anthropological observations resonated in other parts of the global south. In South Africa, George W. Stow, known by the familiar moniker "geologist and ethnologist," was an English maverick who traveled across the Karoo region in the mid-nineteenth century, working in the gold and coal mines and studying the region's geology and ethnology.[173] He subsequently suggested that the Bushmen were the aboriginal inhabitants of South Africa, driven by the "Kaffirs" (a racially offensive term used to refer to the Xhosa [Nguni] tribes) to take refuge in the remote Karoo highlands. Looking to establish the aboriginality of the Bushmen deep in the landscape, Stow found in a bed of sand and marl dug up during mining small stone implements. He suggested that the implements had been deposited when the marl base was forming and had been covered by several layers of subsequent deposits.[174] The hunters who used those implements must have lived in the area when the marl beds were on the surface and the entire region was covered by a "magnificent lake" several thousand years ago.[175] The lakes dried up subsequently and the beds were covered, along with the tools, by other sediments. To Stow, this was evidence that "the Bushman race must have occupied South Africa, continuously, for an enormous period."[176] The Bushman, Stow suggested, had therefore evolved with the landscape.

We now return to the question of why aboriginal people appeared prehistoric. Prehistory carried more intense naturalistic connotations than other categories such as the savage or the barbaric. It was defined by a deep and escapist naturalism, by resorting to a nature that seemed pristine, constant, untainted, and often in the depths of the earth. That nature seemed to be the *only* continuous link for geologists and anthropologists between the present and the deep past. As aboriginality and indigeneity also came to be defined in terms of nature, the people whose lives seemed embedded in the earth were deemed prehistoric. This, in turn, explains why the environment and landscape became so critical to understanding tribal history. In the next chapter, I will explore how specific landscapes defined these debates, with reference to Gondwana.

Conclusion

In 1851, the British archaeologist Charles Thomas Newton declared that the "record of the human past is not all contained in printed books." Beyond texts, he found human history engraved not only in the pyramids of Egypt, the ruins of Assyria, and the marbles of the Parthenon but also in "living" folk-songs and peasant customs and rituals.[177] In the nineteenth century, the question of human antiquity embraced a wide range of intellectual inquiries and traditions. There were indeed collective attempts by European archaeologists, anthropologists, philologists, and geologists to locate the varying traces of human history. At the same time, there was also a linearization of these various traditions into a natural-historical narrative. This does not mean that the "geological" and the "historical" modes of knowing human antiquity were without conflict.[178] Moreover, as we have seen, the geological was not devoid of the historical. Geological readings of stones, bones, and strata continued to draw from textual, social, mythological, and philological contemplations. Yet the geological frame dominated particularly as prehistory with all its naturalistic undertones became the main denominator in deliberations of human antiquity, and subsequently aboriginality. Consequently, within colonial anthropological and geological literature, Indian tribal populations appeared to be remnants of prehistoric races. They were also placed within the narrative of a southern migration of the human race, that is, the first wave of migration out of Africa, settling in remote mountains and valleys and remaining both primitive and prehistoric.

In recent times, the history of early humans has become the domain of genomic research. One of the reasons why such genomic determination of human antiquity became possible was because the deep past of humanity was linked to its contemporary aboriginality. This created the possibility for DNA research to be conducted among living aboriginal or indigenous populations to trace the deep human past.[179] This chapter contributes to the growing realization among historians that questions of human race and origins need to be reconfigured as historical problems, particularly with recent genomic studies of race and human antiquity. Historians have challenged the biological and genomic determination of race and have suggested that cultural assumptions about race were embedded in the Human Genome Project, which mapped the complete DNA sequence of the human genome and identified its component genes between 1990 and 2000.[180] The genomic project is a product of the naturalization of the history of human antiquity that started in the nineteenth century. In a more delinearized reading, for example, the out of Africa thesis might appear to be a secular form of Genesis history, of the dispersal of the original stock of humanity, the kind that Jacob Bryant referred to in his work on the journeys of Noah's sons. Andrew Shryock, Trautmann, and Clive Gamble have made the significant observation that in modern genomic trees, which piece together that nearness of all humans to each other, there is a "distinctly biblical resonance," in which Africa is the "new Eden."[181]

In this chapter I have explored how some of the scientific premises of the deep antiquity of humanity were constructed in the nineteenth century. In India, as well as in Australia, Africa, South America, and even parts of Europe, the question of human antiquity developed in concurrence with anthropological and geological insights. Non-tribal people have given tribal populations various epithets, such as "the aboriginal," "the primitive," *bon manush*, or "the indigenous," each of which is a matter of intellectual and political debates. "Prehistoric" is a slightly later and more specific category, although it combined elements of these other preconceptions. It not only equated these living populations with humans belonging to thousands of years back; it also provided the naturalistic prism through which these humans and subsequently their aboriginality, primitivism, and indigeneity could be seen. Within this frame, certain populations, such as the aboriginal tribes, appeared more "natural" than others.

CHAPTER FIVE

The Other Side of Tethys

Gondwana and the Geologies of Primitivism

The British-born Austrian geologist Eduard Suess coined the term "Gondwanaland" in 1885 to describe the ancient southern landmass that according to him comprised southern Asia and Africa.[1] Later, South America, Australia, and finally Antarctica became part of it as well. It was distinct from the northern landmass, which Suess called Eurasia (later named Laurasia, also known historically as the Angara), which comprised Europe and North America. The Tethys Sea was supposed to have separated the two.[2] Gondwanaland is a metonymical category that is local, global, historic, and prehistoric simultaneously. Suess derived the term from the Gondwana region in central India, which in turn was named after the Gond tribes who lived there, whom British ethnologists believed to be the aboriginal inhabitants of India. Gondwanaland became a global category through the geological imaginations of an ancient landmass that encompassed different parts of the Southern Hemisphere. This prehistoric supercontinent has also led to the identification of modern cultural landscapes. There is a game reserve called Gondwana in South Africa and a rain forest by the same name in Australia, both having derived their identities from the ancient geological landmass. Each in turn also has its distinct history of deep time.

In this final chapter, I turn to the process of how the Gondwana region of India, with its specific geological features and tribal demography, became a landscape of prehistory in the nineteenth century. This process, as has just been indicated, was part of the larger creation of the prehistoric continent of Gondwanaland. The simultaneous making of Gondwana and Gondwanaland reflects the essential divide between Europe and its colonies. It situates Gondwana as a product of nineteenth-century ideas of colonial deep time, where the imperial exploitation of natural resources took place alongside the search for the primal and primitive nature. Primitivism became a pervasive colonial theme in the nineteenth century. It carried meanings of aboriginality, backwardness, and savagery and was used mostly for different colonized populations. In this period, primitivism was also invested with geological primevalism, of nature in its deep prehistoric state. It also referred to colonial landscapes in general. During his travels in South America, Alexander von Humboldt sensed primitivism in the rocks, the mountains, the strata, and even in the climate of the Andes as much as he did among its native inhabitants. He believed that there nature presented itself in a primal state.[3] The association between primal and colonial nature also acquired the geographical motif of "southern-ness," which was manifested in Gondwanaland, in the colonial south of South America, Africa, South Asia, and Australia.

As a geological and anthropological theme, Gondwanaland runs deeper than other tribal-geological terms such as "Ordovician" and "Silurian," which were "invented" and inserted into intellectual discourse by Victorian geologists contingently to resolve purely geological debates and had little anthropological antecedence.[4] Since the anthropological and geological convergence of terms such as "Ordovician" or "Silurian" are yet to be explored, they have remained purely geological categories. In contrast, the British adopted the term "Gondwana" from precolonial Mughal texts, and it became part of a common parlance used by geologists, ethnologists, and archaeologists in their search for the deep and primitive antiquity of India in both its human and geological forms. This convergence of geological and anthropological investigations in central India was underpinned by the colonial conquest of the region. Gondwana was a vast territory of hills and ravines, where ethnological studies of presumed aboriginal tribes and geological studies of the oldest rock formations of the Indian subcontinent shaped ideas of ethnological aboriginality and geological primitivism simultaneously. Since the primitivism of Gondwana was identified in both its landscape and its inhabitants, the interplay of different antiquarian disciplines is evident there. In its different usages in India, "Gondwana" came to represent something that was ancient, primitive, and prehistoric. In the following, I explain why such a composite deepening of time was essential to colonialism.

Conventional histories of Gondwanaland present a relatively linear and Whiggish account of how the idea developed out of nineteenth-century geological exploration in different parts of the world and was gradually assimilated into the continental drift theory in the twentieth century.[5] These narratives revolve around the work of Suess, Wegener, and du Toit.[6] Suess first assembled the ancient supercontinent based on the discovery of the *Glossopteris* fossil flora in the coal and gold mines of peninsular India, South Africa, and Australia. Du Toit and Wegener eventually linked Gondwanaland to theories suggesting that the southern continent drifted apart.

The idea of Gondwanaland as a distinct supercontinent, as we imagine it today, did not emerge until the second half of the twentieth century. Suess's Gondwanaland was formed by land bridges, which he believed were subsequently submerged as the earth cooled and contracted. Wegener challenged this theory in 1912, suggesting that the continental plates had moved, breaking up the ancient supercontinent. This gave Gondwanaland the status of a distinct southern supercontinent. Even then, the theory of continental drift was widely contested, particularly by American geologists, who remained more convinced of species migration across land bridges than of the moving of continents.[7] It was only in the 1950s, when the continental drift theory came into wide acceptance, that the idea of a separate continental Gondwanaland emerged. Throughout the nineteenth and early twentieth centuries, therefore, Gondwanaland had various shapes and meanings and remained the intimate Other of Laurasia, separated by the Tethys (fig. 5.1). It is significant to note that while geologists in the late nineteenth and early twentieth century fiercely debated the precise shape and origin of Gondwanaland, they never doubted its existence or questioned the north-south divide of geohistory. Gondwanaland was as real as the colonial world.

Rather than writing a conventional history of the making of Gondwanaland, which a British-Indian geologist impassively described as a "dim and frozen continental region," I will retrieve this history from geology. In other words, I will distinguish between the historical and the geological imaginations of the deep past. Geology itself is a historical discipline, as far as it is an imagination of the past. However, in its positivist and evolutionary approach to antiquity, it overlooks the different shapes and meanings of Gondwanaland. I deconstruct that narrative first by exploring the origin of Gondwanaland in the British colonization of the densely forested region of Gondwana in central India. I then identify the processes through which Gondwana was identified as a site of deep time.

Pangaea Divided

Gondwanaland suggested a fundamental natural and historical divide between the southern and northern parts of the earth. Suess proposed the name in his *Das Antlitz der Erde* (originally published between 1883 and 1909, translated as *The Face of the Earth*), where he suggested that the conventional and colloquial divides between the Old and New World were false. North America was, in geological terms, an ancient continent. At the same time, the seemingly

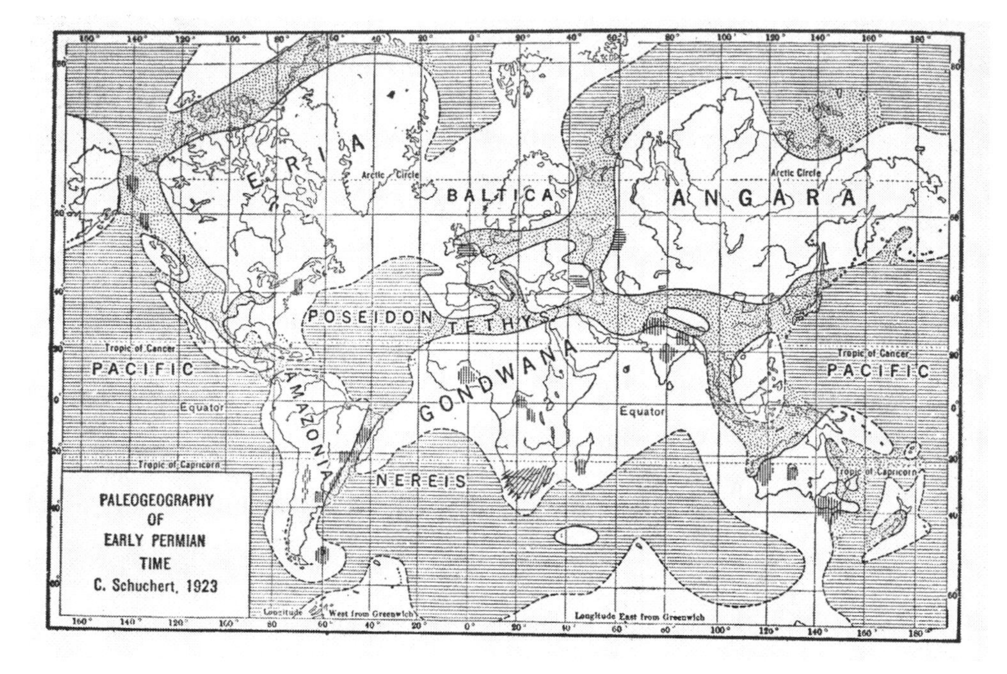

Fig. 5.1. An early twentieth-century depiction of the land bridges of Gondwanaland. Louis Valentine Pirsson, *Introductory Geology*, part 1, *Physical Geology*, by L. V. Pirsson; part 2, *Outlines of Historical Geology*, by Charles Schuchert (New York: J. Wiley & Sons, 1924), part 2, © British Library Board 455.f.3, 564.

"united mass of Asia, Africa, and Europe" was geologically incompatible, as it was made up of several "heterogeneous" regions. Similarly, the Americas were not a homogeneous entity. Suess believed that, geologically, the northern parts of America had much more in common with Europe than with southern America, which shared commonalities with southern Africa and Asia. Having thus reorganized global geography, he initially described the southern landmass as "Indo-Africa," a term that, as we saw in chapter 2, appeared in British Indian geological and zoological discussions in the late nineteenth century.[8] He named the remaining northern part "Eurasia," a region comprising North America and Europe.[9] He identified a "very sharply marked line of separation" between the two landmasses running through the northwest of Africa from the valley of Wadi Draa in Morocco, through the Sahara Desert and the Persian Gulf, to the mouths of the Indus, along the foothills of the Himalayas, into the valley of the Brahmaputra in Assam, finally to the south of Java.[10] Suess wrote, "We call this mass Gondwana-Land, after the ancient

Gondwana flora which is common to all its parts."[11] The proposition was based, among other things, on the discovery of certain plant fossils in distant continents within certain rock formations, which suggested that they were all once connected. The most significant was *Glossopteris indica*, found initially in the coal and gold mines of central India, South Africa, and later in Australia, South America, and eventually in Antarctica.[12] Robert Falcon Scott found the latter during his ill-fated expedition of 1912. The northern landmass was eventually named "Laurasia," a term derived by combining the names of the Laurentian and Eurasian plates of North America and Europe, respectively.

The suggestion of the geological divide between the northern and the southern worlds was preceded by observations of cartographic resemblances between the coastlines of Europe and northern America on the one hand and between that of South America, Africa, peninsular India, and Australia on the other. In the sixteenth century, the Dutch mapmaker Abraham Ortelius suggested that the Americas had broken away from Africa. His propositions were based more on the observation of physical similarities of the coastlines of the continents than on any deep geological knowledge. French geographer Antonio Snider-Pellegrini revived these ideas in the nineteenth century.[13] Wegener adopted them in the early twentieth century with his proposal of the theory of continental drift. He suggested that around 200 million years ago, a single supercontinent, called "Pangæa," began to split apart into Laurasia and Gondwanaland.[14] Wegener, like Abraham Ortelius 300 years before him, had observed the remarkable similarities between the eastern South American and West African coastlines. Wegener then matched the similarities in geological structures and animal and plant fossils in these two regions to suggest that the continents had drifted apart. In the works of du Toit and Wegener, Suess's Gondwanaland became more of an isolated ancient landmass than an amorphous amalgamation of distinct landforms connected by land bridges.

Gondwanaland, as proposed by Suess and defined by Wagoner and du Toit, was Pangaea divided. It was another land, different from Europe and North America. Wegener's Pangaeatic mainland reflected ideas of the biblical paradise as the singular site of the beginning of life. Although he did not refer to biblical metaphors or to theological ideas, the idea of paradise as the source of life had been critical to natural history since the eighteenth century. Carl Linnaeus viewed paradise as a natural-historical category, a central island surrounded by water, from which life originated. The island had all the climates to allow different species to grow. He also supported views of the "whole earth," which resembled Pangaea, before the biblical flood.[15] Ideas of a paradisiac center of the world persisted throughout the nineteenth century as well, including in the work of Humboldt.[16] Although Suess did not refer directly to this paradise, his geological theories carried biblical motifs.[17] He too believed that species had diversified from a central strand. In a moving passage in *Face of the Earth*, he described how species belonging to a single mountain surrounded by the ocean were washed away by successive waves and distributed.[18]

This north-south divide was paradigmatic to modern imaginations of the earth. The South African political leader Jan Smuts referred to Gondwana to stress the centrality of Africa in the southern hemisphere. [19] In scientific terms, the separation between Gondwanaland and Laurasia species can still be identified and is known as the "ancient vicariance." The breakup of Pangaea is believed to have been the single most vital factor in the biological diversification of species.[20] In ethnological terms, Michael Witzel categorizes Afro-Asian cultures and mythologies as "Gondwana mythologies," suggesting a distinct racial and ethnological identity in the southern regions. These Gondwana mythologies are premised on the history of the "southern" human migration of Austric and proto-Australoid races, covering the aboriginal cultures of Australia, the Andaman Islands, Melanesia, and sub-Saharan Africa.[21] These impressions of a divided earth were based on deep colonial encounters. They also conformed to the southern trajectory of aboriginality studied in the previous chapter.[22]

The Forest of the Gonds

Gondwana (or Gond-vana), literally meaning the "forest of the Gonds," became the common term in nineteenth-century India informing descriptions of geological formations with anthropological insights and vice versa. Colonial officials and cartographers adopted the term used in Mughal texts[23] to refer to the vast and largely unmapped territories of central India, wherever settlements of Gond and other tribes were found, from the Vindhya Mountains in the north to the Godavari River in the south, from Malwa in the west to western parts Bengal in the east.[24] A vast trap region, Gondwana is largely an area of deep forests, ravines, hills, and rivulets. The Narmada cuts through deep gorges carving out a rift valley amid the Vindhya and Satpura Ranges.

The colonial geologists who first wrote about Gondwana derived their ideas from specific geological and ethnographical experiences. H. B. Medlicott and W. T. Blanford, in their *Manual of Geology of India*, published in 1879, explained that the term "Gondwana" was derived from the "old name for the countries south of the Narbada valley, which were formerly Gond kingdoms, and now form the Jabalpur, Nágpùr, and Chhattisgarh divisions of the Central Provinces." They simultaneously pointed out that this region was also where "the most complete sequence of the formations constituting the present rock system is to be found." They added in a footnote, "For the information of non-Indian readers, it may be well to add that the Gond is one of the principal Dravidian, or so called aboriginal tribes, who are believed to have inhabited the country before the advent of the Aryan Hindu race."[25] This mixed reference to ethnology and geology is the key to the making of the deep history of Gondwana. Colonial officials, travelers, or missionaries used multiple disciplinary and epistemological metaphors when writing about tribes, fossils, or the landscape and the soil.

For geologists based in India, even in the twentieth century, Gondwana retained the sense of a region geologically very distinct from Europe. The director general of Geological Survey of India, Cyril S. Fox, explained that the "most arresting feature" of the Gondwana system was its "entire disagreement with the recognized sub-divisions of the geological record of Europe." He believed that Gondwana required adopting a timescale different from that of Europe and European geology. He found parallels with another colonial landscape with similar histories of coal and prehistoric geology: the Karoo of South Africa.[26]

From the end of the eighteenth century, British explorers and mineral prospectors had traveled across Gondwana. In 1795, Captain James T. Blunt explored the routes under the instruction of the East India Company. After crossing the Mahanadi River, he entered the forests of Kanker (in the present-day north Bastar region of Chhattisgarh), where the road dwindled to a narrow path through thick bush. Blunt's Hindu guide warned him that this was the "abode only of wild beasts, demons, and the savage Goands."[27] After crossing a low ridge of hills, his party entered a region his guide told him was the land of "the ancient *Rajahs* of *Goandwannah*; and is entirely inhabited by the *Goand* mountaineers."[28] In one of the earliest British references to it, Gondwana appeared as the dense forest of the Gonds.

Gondwana assumed much greater geographical and ethnological proportions, covering different parts of the Central Provinces as colonial encounters with the Gonds and other tribal groups became more extensive. In 1854, the *Gazetteer of the Territories under the Government of the East-India Company* described "Gondwana, or the land of the Gond race" as "an extensive imperfectly-defined tract of Southern India." It covered most of the vast region bordering Bengal in the east and Berar in the west.[29] In 1861, the "territorial division of Gondwána" was regarded as a special part of British India because of its unique geographical, historical, and ethnological characteristics. The colonial authorities named this "central tract of highland and valley, with its unknown history, its unsuspected resources, and its strange world of wild tribes," the "Central Provinces."[30] Even until the 1850s, large parts of the province remained "a *terra incognita* to Englishmen." Thomas Erskine Perry, chief justice of the Supreme Court in Bombay described these "unexplored" "Gondwana highlands and jungles" as an "oasis in our maps."[31] The *Gazetteer* added that these parts of India were comparable to Central Africa. The large forest tracts were supposed to be inhabited by humans who "live in trees, and . . . there were whispers of 'anthropophagi'—naked savages who ate their relations."[32]

Along with these supposedly naked savages, British explorers were struck by the stark marble cliffs in Gondwana, the turbulent Narmada cutting through ancient rocks, and the volcanic eruptions that had provided the black soil vital for growing cotton. The early British explorers and administrators referred repeatedly to the cultural landscape of Gondwana.[33] W. H. Sleeman, of the Bengal army, traveled across the region between 1825 and 1835 and wrote of his experiences, published as *Rambles and Recollections* in 1844. His descriptions of the region and of the Hindu and Gond cultural and religious practices were laced with references to this intensely geological landscape. He noted that according to local beliefs, the basaltic columns were the petrified remains of the Gond bride and groom who had dared to glance at each other too soon before their conjugal union. He visited the spectacular Bhedaghat, where the Narmada cuts through high marble cliffs. He felt the cultural power of these rocks and noted that "every fantastic appearance of the rocks, caused by those great convulsions of nature which have so disturbed the crust of the globe, or by the slow and silent working of the waters, is attributed to the godlike power of those great heroes of Indian romance, and is associated with the recollection of scenes in which they are supposed to have figured."[34] Donald Friell McLeod came to central India in 1830 as a colonial administrator and felt a similar reverence for the ancient landscape. The hills around Seoni (around 100 miles north of Nagpur) reminded him of the Scottish Highlands.[35] His own family belonged to an old Scottish Highland clan known as the McLeods of Assynt, who were believed to be descendants of the Norwegian chief from the ancient kingdom of Sogn in southwestern Norway who settled in Assynt.[36] McLeod described the similar "romantic beauty" of the dense forests, the flow of the Narmada through the marble cliffs, and the stark columnar basalts of Mandleshwar.[37]

In this seemingly primal landscape, British geologists found what they believed to be the oldest rock formations in India, which they named the Vindhyan Mountains, and the oldest fossil-bearing rock series, which they called the "Gondwana series."[38] They also found the Gonds there, who appeared as primal as the landscape itself (fig. 5.2). They seemed synonymous with these hills. Hislop, the Scottish missionary who worked among them, believed that the name "Gond" had emerged from the Telugu word *khond*, or *khund*, which meant "mountain"; thus, the term "Gonds" signified "Hill people." He thought that "no designation could be more appropriate to the localities which the majority of them inhabit."[39] Colonial explorers sensed primitivism in this

Fig. 5.2. Gonds of central India. George Smith, *Stephen Hislop; Pioneer Missionary and Naturalist in Central India from 1844–1863* (London: John Murray, 1888), 221. From the National Library of Scotland, used under the Creative Commons Attribution (CC BY) 4.0 license, https://creativecommons.org/licenses/by/4.0/.

seemingly changeless, vast, and overbearing landscape and its inhabitants. Ethnologists, administrators, geologists, and missionaries found in the landscape and among its inhabitants a primeval purity of life that also appeared to be rare and fragile in the nineteenth century.

Ethnologies of Conquest

Tribal ethnology in India developed along with the British conquest of central India. With the fall of the Marathas in 1818, large parts of central and western India came into British possession. Richard Jenkins was the resident of the EIC in the Maratha capital of Nagpur when he advocated the company's aggressive policies against the Marathas and the destruction of Maratha power in 1810. The possibilities of tribal ethnology became simultaneously evident to him. Soon after he arrived in Nagpur in 1807, Jenkins wrote to Mountstuart Elphinstone, the governor of Bombay, that he planned to collect information about the "ancient and the modern states of this country," particularly through the history of the Gonds. He believed that knowledge of the history and language of the largest tribal population of the region would provide the British with an intimate understanding of the people and their land. He soon started his "Gondee inquiry."[40]

Ethnological studies among Gonds and other tribes were defined by the need to understand the nature of tribal resistance to British occupation, which continued even after the fall of the Marathas. In 1819, Jenkins organized military forces against rebellious Gond chiefs.[41] British troops invaded Gond villages, confiscated their cattle and "other articles of plunder," destroyed their grain, and burned down their villages.[42] Jenkins sought to demonstrate to the Gonds, as he wrote to Governor General Warren Hastings, the might of the British forces, "the irresistible power which they had provoked" through their resistance.[43] Most of the Gond chiefs surrendered subsequently. The capture of one of the most prominent rebellious Gond leaders, Cheyne Sah, in the dense forests of Gondwana in 1820 put an end to the immediate resistance. Jenkins reported to Hastings, "No one [is] daring to resist, or thinking himself safe in his most inaccessible retreat, after that powerful, and by then dreaded chieftain had been dragged from a fastness which beggars all descriptions in its wild and terrific situation in the depths of an almost bottomless abyss."[44]

Jenkins subsequently sought to pacify the Gonds and bring "tranquillity" to the region.[45] To him, tranquility meant "civilizing them." Jenkins insisted that the Gond chiefs pay annual visits to Nagpur, where they could learn about

British values of civility and kindness.[46] Brigadier General John Malcolm, who in 1818 was appointed by the Marquis of Hastings in military and political charge of central India, suggested that Gonds should be encouraged to cultivate the land on both banks of the Narmada to encourage them to give up plundering.[47]

This vision of pacifying the tribes though colonial modes of settlement ultimately led to further displacement from their land. In subsequent decades, with the formation of the Central Provinces in 1861, the British rulers, particularly Chief Commissioner Richard Temple, adopted a land revenue arrangement with the tenants, known as the *malguzari* system.[48] It was based on a romantic notion of independent and autonomous village systems that Temple cherished. However, this colonial land reform system ultimately empowered the Hindu peasants and further alienated the rights of the tribal population, particularly the Gonds. It encouraged the tribal populations to convert several forest tracts into agricultural lands, which led non-tribal populations to settle in these regions and evict them.[49]

Another strategy of ensuring peace was to trace an alternative racial and genealogical lineage of the Gonds. The British saw the Gond rebellion against their occupation as a product of the political and social contamination of the tribes, by which their leaders such as Sah had adopted the Maratha forms of political leadership and resistance to the British. The early ethnological studies sought to establish the primary features of Gond ethnology: a study of their racial ancestry through the study of their language, establishing their racial distinction from Hindus and situating them as the aboriginal races of India.

British military officers or administrators who followed the colonization of central India conducted the initial studies of Gond history. Captain C. W. Montgomerie, who was responsible for drawing up the settlement with Gond chiefs once the rebellions were crushed, produced a genealogy of their families, their traditional rights and powers, and their political alliances with the Marathas, Muslims, and Brahmins.[50] As he compiled the detailed family history of Cheyne Sah, he learned about the Gonds' long history of struggle against Muslim rulers.[51] John Malcolm, who had suggested that Gonds should be encouraged to cultivate the land, also conducted detailed investigations into Gond genealogies. He found that the Marathas were never able to establish "anything like complete rule over these countries" and the Gonds had remained outside the major state formations in the region.[52] Malcolm believed that Gonds were distinct from the Hindus and comprised different ethnicities, including other tribes who lived among them, such as the Bhils and the Bagias.[53]

The broad conclusions of these studies were that although ethnically a distinct race, politically and culturally the Gonds had become part of the feudal landscape of precolonial India. Jenkins concluded in his "Gondee inquiry" that the Gonds, particularly the ruling "Rajahs" among them, considered themselves of Kshatriya (Coeetoor) or Rajput descent. He also noted that the Gonds were described in historical Hindu texts as a "powerful nation," which led him to suggest that they were essentially the primitive races of India, who had been conquered by the Hindus and driven to the hills and forests where they now lived. There the rulers among them adopted the feudal system and formed revenue arrangements with the Mughals and Marathas and thus became part of the caste-agrarian landscape of India.[54] Temple, who arrived as the acting chief commissioner of the Central Provinces in 1862 and wrote on Gond ethnology, commented on this hybridity: "There are . . . in the present age . . . three kinds of Gonds, namely, the aboriginal Gonds, the Hindu Gonds, and the few Mussulman Gonds."[55]

Therefore, in the early ethnological studies, the ethnic, political, and cultural contamination of the Gonds appeared as the key theme of the study of tribal aboriginality of India. The perceived mixing of the tribal population with caste Hindus that troubled the British invoked references to the geology of the landscape and coincided with geologists discovering the oldest sandstone formations of the subcontinent along the same Narmada valley.[56] Colonial ethnologists and geologists referred to this geological antiquity to stress the aboriginality of the Gonds. As James Forsyth, another colonial administrator who traveled across the central Indian highlands in the 1860s in search of minerals, used geological motifs to address this problem of ethnological hybridity:

> Something has already been said regarding the intermixture of Hindu blood, manners, and religion, that has taken place among the aboriginal races of Central India. Were this an isolated event in the ethnical history of the country it would possess a comparatively feeble interest. Its high importance lies in its furnishing us with a living example of a process which has, as already suggested, played an important part in the development of the races which compose the mass of modern Hinduism. It is the uppermost and most accessible stratum of a geological series of untold antiquity; and, as the geologist interprets ancient formations by the analogy of the processes he sees still going on around him, so it may be that some light may be thrown on the construction of modern Hinduism by the process of transformation which is here going on before our eyes.[57]

In this seemingly ethnological account of the tribes, geological metaphors linked the hybrid reality of the present with the deep past. Forsyth saw the tribal history of Gondwana as a part of its geological narrative. For both the ethnologist and the geologist, the Gondwana they saw held secrets of deep time. Seeking to sift out the Brahminical, Sanskritic, and Puranic influences both in its humans and its nature, these ethnographers and geologists were searching for an India that was essential, pure, and primitive. The peculiar landscape of central India provided the perfect setting to discover the true ethnicity of the Indian tribes. Although in their current state Gonds did not appear to be the most primitive race, the geology of Gondwana provided the clue to this search for primordial antiquity by colonial ethnologists, geologists, and missionaries. McLeod believed that this powerful landscape had shaped human life in this region in a primal way: "It would seem that along this belt has raged the principal force of those subterranean fires which in former ages have heaved up the central portion of Hindustan, and introduced so much of variety in its aspect, its temperature, its productive powers, and, most striking perhaps of all, in its inhabitants, language, and manners."[58] Gonds in their primitive form appeared to be part of this landscape. To Jenkins, the "jungles and mountains" of Gondwana were the "true scene for discovering the 'naked' manners of the Gonds."[59]

Sumit Guha has suggested "strata" as the common metaphor in this alignment between geology and ethnology. Nineteenth-century European anthropologists, he argues, by taking "tribe" and "caste" as timeless and unchanging categories marked by "continuity of racial descent," sought to demonstrate that "South Asians, like other unprogressive people, did not change—they merely accumulated, with the latest addition to the population overlaying its predecessor, much as geological strata did."[60] This association between tribes and strata is useful since, in India, the social status of tribes and castes appears to be vertically layered. Strata, since William Smith's observation, have also been a fundamental feature of geohistory. However, they can be misleading as well. Geologically, particularly in Gondwana, strata were not always vertically layered. As we saw in the previous chapter, different beds, such as Pliocene and contemporary ones, appeared to be *juxtaposed* rather than layered.[61] Second, colonial ethnologists did not always see a layered pattern of human settlement or evolution in India. Neither did they suggest that tribal communities were changeless. Their hybridity invoked alternative geological references. Hislop traveled across Gondwana, preaching the Bible among the Gonds and studying them closely to trace their "pure" state untainted by Brahminical influence.

His biographer, George Smith, wrote, using a geological metaphor about his desire to identify and preserve the pure Gond, "Hislop's heart yearned, and he laboured incessantly to save the Gonds from that Hinduism to which, but for Christian missions, they must inevitably be attracted. It is still a fact of powerful significance for Christendom, as it was for him, that so many millions of simple nature or demon-worshippers are untouched by the *fossilising influence* of the Brahmanical and Puranic systems."[62] Therefore, it was the Gonds' fossilization into the Hindu pantheon both ethnically and politically that troubled colonial ethnologists. The need was to sort the primitive tribes from the Hindus and locate them in the deep history of Gondwana.

This particular synergy between geology and anthropology raises a wider question about habitat or ecology that dominates modern scholarship on tribal history and society in India. Scholars have regularly referred to the geohistorical settings of tribal lifestyle and history as a key element of understanding tribal life.[63] André Béteille has argued that landscape is one of the main denominators of tribal history in India.[64] Such a suggestion derives from colonial ethnographic accounts, which repeatedly identified Gonds with their habitat, referring to them as "mountaineers" or "hill people." For example, to Gustav Oppert, the crucial aspect of Indian aboriginality was that these people lived in hills and forests, away from the fertile plains.[65]

This raises the question, should we read Gond history as part of the geological setting of their habitat, or should we analyze how in fact these links between them and the landscape were constructed? Sumit Guha has examined the historical relationships between ethnicity and ecology in South Asia.[66] His primary concern is with the assumption of changelessness, both of indigenous communities as well as of their relationship with their habitats, or as he puts it, the idea of "relic populations" living in "relic forests."[67] To challenge this assumption, he problematizes the question of tribe as well as that of indigeneity. In a history spanning almost 800 years, Guha establishes the dynamic nature of the interaction among communities living in the riverine plains, forests, hills, and deserts of South Asia with their ecology. He has shown how human action, state policies, and economic processes altered the environment as well as the lives of those communities.

Ajay Skaria, in his innovative study of the Bhils in the Dangs region of western India, in contrast, has produced a more political reading of the significance of tribal habitat. He suggests that the term *jangal* (literally "forest or wilderness"[68]), as used by the Bhils in precolonial India, occupied a very different epistemological space than it did in colonial usage. *Jangal* meant not necessarily forests and mountains but any region lying outside strong state control. Skaria sees the "Jangal polity" as a space that was agnostic and even antagonistic to the bureaucratic and commercial mechanisms of the plains societies.[69] Here Skaria anticipates some of James C. Scott's arguments about communities living in the anarchist and nonstate space of the "Zomia."[70]

These are sophisticated analyses, and they challenge the easy synonymy between tribal history and ecology. Yet while they highlight the dynamic and political nature of the relationship between tribes and their ecology, they do not problematize how and why habitat became such a fundamental and unique feature of tribal history. The answer lies in identifying how the two processes, the methodological convergences between anthropology and geology and the unfolding of the tribe-landscape synergy, developed simultaneously. Primitivism in Gondwana was never simply an ethnological problem; it was deeply ingrained with geological meanings. This particular association has informed modern scholarship. It, therefore, remains important to examine the mutually constitutive relationship between the terrain and the tribe that emerged in the nineteenth century as evident in the concurrent making of Gondwana as an anthropological and geological space. I will explore this problem by analyzing a specific convergence, that of colonial and tribal geohistories of creation and deep time. In the next two sections, I turn to the colonial discovery of the geological past of Gondwana and its alignment with Gond geomyths of their homeland.

The Prehistoric Forest of the Gonds

Since the early nineteenth century, British geologists, missionaries, ethnologists, and colonial administrators looked for traces of prehistoric Gondwana. They marveled at this powerful landscape and speculated how it was created by the action of water and fire, like McLeod, who had sensed that subterranean fires had shaped the landscape and human life there. They were fascinated by the possibilities of finding the history of the origin of life there. The most extensive and striking geological features of western-central India were the traps, commonly known as the "Deccan traps." The term "trap" was adopted from the Swedish word *trapp*, for the volcanic rocks' resemblance to stairs. The term "Deccan" derived from the Hindi/Prakrit word *dakhshin*, or *dakkhin*, which means "south," had been used by the Portuguese since the sixteenth century to refer to the political states of west-central India. Geographically, the Deccan refers to the central tablelands south of the Narmada. The Deccan traps cover an area of around 200,000 square miles, comprising volcanic rocks and enormous stretches of horizontal layers of basaltic rocks, separated into beds with flat-topped hills.[71] It is one of the largest volcanic provinces in the world. From the 1820s, British explorers and administrators in central India, such as Voysey, Malcolmson, Thomas J. Newbold, Charles Grant, H. J. Carter, Hislop, and the (H. B. and J. G.) Medlicott and Blanford brothers described the geology of the traps.[72]

Within these traps, several explorers found beds of freshwater fossils. Voysey was the first to analyze the fossil shells, which were presumed to be the earliest animal remains found in the Indian subcontinent, in the layers of the traps and commented, "It is a remarkable fact, that the only remains of animals hitherto discovered in India, should be found in traps." He found that some kind of pressure had flattened most of them and concluded that some "sub-marine or sub-aqueous volcanoes" had deposited these fossils.[73] Others found similar shells elsewhere in the trap region.[74] These traps and their freshwater fossils suggested that the entire region had been covered in water in prehistoric times. It also suggested that these shells were the earliest forms of life in the subcontinent.

In the 1850s, Hastings Fraser, the British minister at the court of the nizam of Hyderabad, found a specimen of a fossilized fish with impressions of leaves on it in the traps, which also supported theories regarding the prehistoric submersion of the region. William H. Sykes, who had served in India as an army officer, received the fossil fish from Fraser and enthusiastically noted that the discovery was "a novelty necessary of great interest, as indicative of the former submerged state of the peninsula of India."[75] John Malcolmson found fossils on the eastern edge of what was known as the "great basaltic district." He studied the freshwater fossils in the Lonar Lake, which was formed in a crater at the heart of the Deccan trap region. He also analyzed the fossilized shells Sykes collected near the western edge of the Deccan and the ones William H. Benson gathered from Bengal.[76] He believed that these shells must have lived in water, of which there was no trace in the Deccan region. Several eruptions of lava and rocks had disturbed the strata, and the fossils came "from the same secondary rocks being broken up and altered."[77]

Charles Grant, a commissioner of Central Provinces, believed that these basaltic beds pointed to the presence of small lakes of different sizes, formed in between successive stages of the outpouring of volcanic lava.[78] Sleeman had found freshwater fossils in the hills around the Narmada valley between 1828 and 1830 when he traveled in the region. He had also found fossils of several small trees with roots, trunks, and branches, "all entire, and beautifully petrified" along with some fossil bones of animals.[79] In the city of Saugor, which was situated in a flat part of the Vindhyan Range, he found a "grove of silicified palm trees" within a mile of the British cantonment, excavated when the company army cut a road through the forest.[80] He too believed that the traps were submerged in either the sea or a huge lake. In the step-like walls of the sandstone hills, Sleeman saw "marks of the *ripple* of the sea."[81] He wrote that the basaltic top of the hills, which were raised above the water of the prehistoric lake, became the habitation of early vegetable and animal life, traces of which were found as fossilized trees and shells. The *Gazetteer of Central Provinces* also confirmed that these fresh-water fossils in the traps pointed to their origin in lakes and pools that were trapped during successive flows of lava that now covered the entire volcanic region.[82] A consensus gradually emerged among these explorers about the previous aquatic submersion of peninsular India.

Hislop's initial interest in the natural history of Gondwana began as he started collecting insects for his brother Robert, an entomologist back in Britain.[83] In the process, he came across fossilized insects in the traps. Along with his fellow missionary Robert Hunter, he discovered fossilized footmarks of reptiles, small freshwater animals, vegetable remains, scales and jaws of fish, and the entire head of a saurian in the neighboring regions.[84] They also found fossilized fruits and crabs.[85] He had trained his trusted local convert, Vira (Virapa), to collect the fossils.[86] In 1851, while Hislop and Hunter were teaching tribal students at their missionary school, they noticed that the tablets of reddish sandstone that the children used as slates bore imprints of fossil plants. They traced the stones to the quarry from which the children had obtained them and discovered abundant stones with vegetable fossils. They soon visited Sillewada, north of Nagpur, and found "rich and most beautiful specimens of *Glossopteris*" in the local sandstone.[87]

At Koradi, 7 miles north of Nagpur, they found freshwater fossils, fossilized "tracks of worms," and numerous fossilized fish in the traps.[88] They also received freshwater shells from surgeon J. Miller (of the Madras native infantry) and Richard H. Sankey (of the Royal [Madras] Engineers), who found them around Nagpur. Sankey, along with the zoologist and ornithologist Thomas C. Jerdon, had also visited the village of Machagora, around 100 miles north of Nagpur, where they found several well-preserved shells and specimens of wood and fossil shellfish.[89]

Hislop compared these fossils found around Nagpur with those found in other parts of the central Indian plateau. In the 1850s, he sent some of his own fossils to Thomas Oldham, the first director of the GSI, who was overseeing the mapping of the coal-bearing strata in the eastern parts of central India. Oldham confirmed that Hislop had been working on the "continuation of the same bed."[90] Oldham even sent him drawings of the shells and lithographs of some of the plant fossils from western Bengal.[91] These suggested to Hislop that the entire central region of India had similar geological features and origin.

Hislop synthesized these discoveries of fossilized ferns, trees, fruits, fish, and shells into a geological description of the prehistoric forest of Gondwana and the origin of primal life there in two papers he wrote (one jointly with Rev. Hunter) on the geology of central India.[92] He suggested that in the prehistoric period, Gondwana was covered in "an old-world lake."[93] This large freshwater lake covered the entirety of central India, stretching southward to the peninsula, into Bengal to the east and a narrow channel to the sea to the north and west. Organic life appeared for the first time in this lake in peninsular India.[94] Deriving from the various discoveries of shells, plants, and tracks of insects, Hislop described life in this vast marshy land: "On the shores of this lake earthworms crawled, and small reptiles (frogs) crept over the soft mud. In its pools sported flocks of little Entomostracans, resembling the modern *Estheria* [shrimp], mingled with which were Ganoid fishes and Labyrinthodonts [an extinct amphibian]."[95] Hislop's depiction of the prehistoric forest of Gondwana resembled later portrayals of the "coal swamps" of the Southern Hemisphere, the prehistoric forests of the Carboniferous and Permian periods, the decaying plant biomass of which gradually formed coal.[96]

Hislop described the dry land around this vast prehistoric lake as covered in the dense vegetation of ferns like the *Glossopteris,*[97] and "low-growing plants with grooved and jointed stems inhabited the marshes; and Conifers and other Dicotyledonous trees, with Palms, raised their heads aloft." The subterranean volcanic outbursts shattered the top layer of soil, and the extensive explosion of granite elevated the bed of the lake, leaving it dry. Another depression followed, and the region was once again covered by a vast lake. This lake, which "furnished an abode to its peculiar living creatures and plants," was once again covered by a major outpouring of lava that filled up its bed and left a greater part of central and western India "a dreary waste of lava." This lifeless landscape was broken up by a second eruption of lava, which led to the deposits of black soil in the Deccan.[98]

Regur, the characteristic black soil of this region, was an important part of this emerging prehistoric imagination of Gondwana. Often known as "cotton soil" for its suitability to the cultivation of cotton, vital to the British imperial economy in the nineteenth century, the term *regur* was derived from the

Telugu word *rēgaḍa* (black soil). A rich, dark, loamy soil, it was known for its moisture retention. It was extremely rich in organic matter and was believed to have produced crops for 2,000 years without manure or irrigation.[99] Geologists investigated its characteristics in detail.[100] They found *regur* wherever there was the volcanic trap throughout the Deccan and the Narmada valley.[101] Captain Newbold, assistant commissioner for Kurnool (on the southeastern edge of the Deccan), believed that almost one-third of southern India was covered by *regur*.[102] Geologists believed that *regur* was "disintegrated trap" that could explain the history of these traps.[103] W. T. Blanford believed that the black soil in different parts of central and western India showed evidence of every stage of decomposition from hard basalt to *regur*.[104]

This was not merely a prehistoric inquiry. The soil was critical to the colonial economy in the newly acquired territory. The colonial government consulted Hislop about the distribution and special characteristics of *regur* while initiating the *malguzari* land settlements in the Central Provinces. While drawing up the details of the settlement, the authorities valued Hislop's acquaintance with the local peasants and his knowledge of the geology of the region for his analysis of the *regur*.[105] Hislop believed that the secret of its high fertility was in the impregnation of the soil with organic matter, probably the remains of the dense prehistoric forest.[106] Blanford agreed: "Everything observed in Western India tends to confirm this view."[107]

Hislop's theory of the formation of *regur* became an integral part of theories of the emergence of life, even humans, in the Deccan. Foote found that the early human tools discovered in the region had been carved out of this landscape. While identifying the main prehistoric sites in southern India, he noted that the prehistoric axes and *celts* "without a single exception" were made of "trappoid rocks," or those belonging to the Deccan traps.[108] He suggested that early humans had lived in the Deccan region, which was then covered in dense forests. Drawing from "Hislop's regur formation theory," as he referred to it, Foote suggested that the trees had died and led to the spread of forest humus, which formed the black soil.[109] The theory, he stressed, was that the destruction of the luxuriant forests by fires led to the formation of the rich humus, which entrapped organic matter. The black soil absorbed plants, insects, and entrapped small animals from the surface, which decomposed and sustained its fertility. The colonial cultivation of cotton thrived in this soil, which had also supported human life for thousands of years. They found that these geological processes were, in fact, taking place in front of them. Foote once observed the violent enactments of life and death in the black soil:

> I experienced two such tempests [storms and floods] while working in the Bellary District, so can speak of actually seeing their effects. Lizards and other small animals, and immense numbers of insects of many varieties, and vast quantities of vegetable matter were washed down into the gaping fissures, which were soon filled up and the washed in objects securely buried.[110]

Hislop met his end in these tempestuous waters and soil. In September 1863, on his way back from his archaeological expedition in Takalghat, he was separated from his horse in the darkness while crossing a swollen stream during the rains and drowned. His body was found next morning in the riverbed, clutching at the turf of the riverbank. In one pocket, he had his Bible and in another a few prehistoric items from his excavations of that day.[111] The deep earth of Gondwana had claimed him.

Hislop's theory of the formation of Gondwana and of life in the Indian subcontinent was gradually absorbed within the emerging theory of the intercontinental links between India and Africa and the idea of a greater Gondwana across the southern oceans. George Smith, Hislop's biographer and himself a geographer, connected the theory of the emergence of life in central India to the emerging global history of Gondwanaland. To him, the traps of Nagpur were part of the vast continent stretching from Madagascar and South Africa to Malaysia and Australia. In those eruptions and overflows lay "the history of the preparation of the world for man."[112] This basaltic soil, once the water drained off, became ideally habitable for humans and thereby allowed "teeming millions, who live upon its millets, [to] export its wheat, and clothe themselves and half the world with its cotton."[113] In one instance, George Smith connected the prehistoric formation of Gondwana to its colonial present.

Several geologists who had worked in central India and then traveled to the northern and eastern parts of Africa on imperial duties observed the geological similarities between the two regions. Newbold, who had made the original analyses of *regur* before Hislop, explored Egypt for two years, from 1845 to 1847, after leaving India. He observed that the black soil along the Nile was very similar to *regur*, and it too was used for cotton cultivation.[114] In 1867, W. T. Blanford traveled to Abyssinia as part of the British expedition led by Lord Napier against the Ethiopian emperor Tewodros II. He stayed on after the forces left to continue his geological and zoological investigations. He too found that in parts of Abyssinia, the black soil closely resembled the *regur* of central and western India. It was also similarly fertile and suitable for growing crops.[115] Blanford collected several fossils from the Abyssinian beds, which he sent to the British

Museum.[116] There he found trap-like formations and agreed that the secret behind these traps and their remains of organic life was "aqueous denudation."[117] The traps were similar to those of the Deccan, and he speculated that the central Indian and Abyssinian highlands were part of the same formation:

> What connexion exists between the various upper secondary or lower tertiary bedded traps of South-western Asia and Eastern Africa? Should they be proved to have been formerly connected, and to be portions of the same great ancient volcanic region, an idea which seems by no means improbable, their study will become one of very great interest as connected with the geological history of the earth's surface.[118]

The connections imagined were not just geological. Blanford found remarkable similarities between African and Indian fauna. He suggested that while the fauna of eastern India resembled that of Malaysia, the fauna on the west coast and in central India had distinct similarities to that in eastern Africa.[119] It is important to remember that at this time, the Indo-African landmass was still conceived in terms of land bridges that allowed for the imagined migration of species. W. T. Blanford was the first geologist to propose a southern continent as he noted the presence of *Glossopteris* in Africa, India, and Australia.[120] In 1875, his brother, Henry F. Blanford, made one of the earliest geological references to the "Indo-Oceanic" continent. He suggested that the southern and central parts of India had little geological structure in common with the mountains and Gangetic Plain of the north. The Vindhyas, the oldest mountain range in India and rich in paleobotany, particularly the *Glossopteris,* along with the rest of the peninsular Indian rock formations, resembled those of the Wollongong and Newcastle beds in Australia as well as those of the Karoo region of South Africa.[121] From the west coast of India to the Seychelles, Madagascar, and Mauritius extended a line of coral atolls and banks, including the submerged Adas bank and the islands of Lakshadweep and the Maldives, that indicated the existence of a submerged mountain range or a chain of mountains.[122] Suess later described it as the "Indo-African tableland."[123] In 1890, at a time when Suess was making his propositions about Gondwanaland, W. T. Blanford commented on the "extraordinary" similarities in the strata of Australia, India, and South Africa. To him, they appeared to share a sequence of boulders at the bottom, then beds with coal, then coarse sandstone, and then clay.[124]

Around the same time, geologists in South Africa likewise found that the geological evolution of the Karoo was remarkably similar to that of the Indian

Gondwana. Stow published a long report on the geological formation of South Africa in 1871, where he drew upon evidence from India and Australia to support the theory of the former existence of a great "Southern Continent."[125] He too suggested a previous submersion of the entire region. These lakes, he believed, did not belong to the sea, as the coastal regions were relatively free from the signs of submersion. Stow linked the freshwater lakes and water bodies in the Karoo to glaciation: the glaciers melted as the climate changed and the region became warmer.[126] He connected this observation with ideas of the ancient southern continent based on the similarities of species, including the *Glossopteris*, in peninsular India, Australia, and South Africa, hinting at their common origin.[127] The rise of the sea placed an impassable barrier between species that had earlier traveled across the connected continents. The coral reefs across the Indian Ocean, Stow believed, were "crowning the tops of the ancient mountains of a subsiding continent."[128]

Suess incorporated the different descriptions of the inundation of Gondwana and the Karoo into his theory of Gondwanaland. He commented on the similarities of the structures of these two connected regions and on the series of similar plants and peculiar reptiles that flourished in both. According to him, the encroachment of the ocean into the interior table mountains had led to the inundation observed in the fossils of these two regions. Islands such as Madagascar and the Seychelles stood out "of the abyss of the ocean."[129]

Gond Myths of Deep Time

These seemingly geological propositions of the formation of Gondwana need to be situated within their natural-historical setting, which clarifies how the connections between tribal history and the deep history of their habitat were established. The Gond myths of creation, as recorded in the nineteenth century, were remarkably similar to those of the geological history of Gondwana proposed around the same time. Gond priests had sung songs of the creation of the Gond race, commonly known as the Legends of Lingo, for generations at marriages and festivals. Hislop was the first European to record this folklore, in a manuscript originally composed in Gondi, which he had partially translated into English before his premature death. Richard Temple translated, edited, and annotated it more fully. The *pradhan*, or the Gond priest who recited these songs to Hislop, was based at the British headquarters in Nagpur, where he was taught Hindi and Marathi. According to Temple, the *pradhan* translated some of the Gondi words into Hindi and Marathi, from which they were then translated into English, presumably after Hislop's death.[130] The stories narrate the creation of the world and of the Gonds from an almost animalistic race to a civilized people under their savior, Lingo.

Several ethnographers and missionaries collected these stories subsequently from various other sources, and different versions of the myth are available. The common theme is that the Gonds were born from the black soil, the water bodies, and the trees of Gondwana and lived as "savages." The ancient forests of Gondwana appeared synonymous with the Gonds, as Temple's translation of the stories narrated:

> Hither and thither all the Gonds were scattered in the jungle.
> Places, hills, and valleys were filled with these Gonds.
> Even trees had their Gonds.[131]

Lingo appeared and rescued them from their savage ways, and they gradually took to cultivating paddy in the black soil.[132] This story of the redemption of the Gonds from savagery to civilization had different cultural and religious interspersions, often with Hindu religious icons. Chatterton narrated how the primitive Gonds incurred the wrath of the Hindu god Mahadeva for their savage ways and were "embowelled in the earth." Lingo then appeared as their savior and formed their distinctive lineage.[133] According to another version, in the beginning, there was water everywhere, and Lingo was born in a lotus leaf and lived alone. Then the crab and the worm appeared. The crab dived to the bottom of the sea, where it found the earthworm. The earthworm brought up the earth with its mouth, which the crab brought to Lingo, who then scattered it over the sea, and patches of land appeared. Subsequently, black soil appeared, and it began to rain, and it rained incessantly for three days. The rivers and streams filled up, and the fields became green with paddy. The land of the Gonds was thus created.[134]

These were not unique instances of the collection of aboriginal creation myths in the nineteenth century. Several ethnologists, missionaries, and explorers collected similar stories from aboriginal populations in South Africa, Australia, and South America. They were drawn to these stories because they appeared to hold clues to the antiquity not just of these people but also of their lands. In South Africa, Wilhelm Bleek collected Bushmen folklore, through which he situated the San race within the basic features of their natural world—the origin of stars, the wildlife—establishing them as the "first peoples" of the Kalahari.[135] Bleek, Stow, and others also used the San myths they recorded to interpret their prehistoric cave paintings.[136] Joseph Orpen, a British administrator of southern Africa, recorded myths from his San guide, named Qing, in the Maloti Mountains (in present-day Lesotho) in the 1870s. These stories have become, according to Rachel King, the "bedrock" of San cosmology.[137] Skaria has studied the Bhil myths of creation called *goths*, which narrate similar stories of their ancestors' encounters with spirits and nature and how the Dangs region of Western India was made suitable for human habitation.[138] To the nineteenth- and early twentieth-century ethnologists, administrators, missionaries, and geologists who collected them, these stories served the dual purposes of anthropology and geology; they were stories of the origin of the Gonds, the Sans, or the Bhils, and of the prehistory of these regions.

Scholars have deliberated extensively upon the significance of these aboriginal myths, the politics of their chronicling, and their impact on our understanding of aboriginality and tribal ecology.[139] Eduardo Viveiros de Castro's study of Amerindian myths, as a form of "multinaturalism," is one of the most significant interventions in indigenous traditions of nature and culture.[140] Castro urges us to appreciate aboriginal myths within their own cultural paradigm. For example, he provides a complex analysis of the Amerindian understanding of humanity and animality that challenges Western distinctions between nature and culture.[141]

While that is an important perspective, the significance of indigenous creation myths is not only in the appreciation of the alternative naturalisms that they engendered. As we have seen throughout this book, Western perceptions of nature, particularly about the relationship between nature and culture or even between humans and nonhumans, were not monolithic. Moreover, it is difficult to identify any clear or sustained line of divide between aboriginal/tribal and Western naturalisms. As we shall see, Gond cosmologies of creation were co-opted within European naturalized antiquity, which was, in turn, absorbed within Gond cultural and political identities. The process was multilayered and reductionist at the same time. It incorporated diverse conceptions of nature from geological, anthropological, and archaeological insights and linearized them into the modern imagination of the deep past within both Western and non-Western traditions. Therefore, it is the *naturalism* within which these myths were incorporated into the colonial archives that needs to be problematized. A critical analysis of this naturalism allows us to identify the links between the geological and mythological narratives of the prehistory of Gondwana. In turn, it also helps us to understand how tribal history and the ecology of the Gonds' land appear inseparable from each other.

While documenting both narratives of deep time, however, neither Hislop, Forsyth, Chatterton, nor Temple acknowledged the similarities between the geological and the Gond accounts of the creation of Gondwana. The resemblance was not necessarily a conscious one. The significant point is that in both depictions, documented within the colonial archives around the same time, deep naturalism defined Gondwana, situating the Gonds as geological entities embedded in the soil, rocks, and forests of Gondwana. Chatterton noted that the Legends of Lingo was in "sympathy with the jungle." These myths of creation also resonated with Chatterton's own Christian faith, where Lingo as the savior of the Gonds appeared as the savage Christ. In his words, Lingo was an "emancipator," "a wonderfully noble character," and "a soul naturally Christian."[142]

Rachel King, in her study of San myths, has provided another clue to understanding the politics of these myths' collection in the nineteenth century and thereby why tribal habitat and tribal history became critical to colonial anthropology. She suggests that Orpen's collection of San myths needs to be seen in the context of the changing territoriality of southern Africa, particularly Orpen's concerns over the steady loss of the Lesotho chief Moshoeshoe's territories to illegal encroachment by white settlers. The San stories that Orpen collected were intended to reentrench San rights to the land. Therefore, it is useful to see these myths and their naturalism within the politics of the nineteenth-century encroachment on tribal lands and on their pasts. There is an inherent duality here. On the one hand was the colonial mining, settlement, and agricultural expansion that led to the loss of tribal land. On the other was a yearning for a pristine past, evident in the study of Gond and San genealogy and the prehistory of the land. The disciplines of geology and anthropology, in reference to colonial mining, agriculture, and the search for the deep past, engendered this duality. This is particularly evident in a description of the Gonds as mining laborers. On a visit to the Nerbudda Coal and Iron Company in Mohapani, Forsyth found that the Gonds were regarded as the best laborers to delve into the dark depths of Gondwana to extract coal. He believed their courage came from their religion. This was an instance when ethnology and geology came together in the depths of Gondwana:

> The universal pantheism of the Gond stands him in good stead on such occasions. From his cradle he has looked on every rock, stream, and cavern as tenanted by its peculiar spirit, whom it is only needful to propitiate in a simple fashion to make all safe. So he just touches with vermilion the rock he is about to blow into a thousand fragments with a keg of powder, lays before it a handful of rice and a nutshell full of Mhowa spirit, and lo! the god of the coal-mine is sufficiently satisfied to permit his simple worshipper to new way as he pleases

at his residence. If utility is, as some have thought, a good quality in religions, surely we have it in perfection in a pliable belief like this.[143]

Gond myths of antiquity were thus naturalized within the colonial political economy. On the one hand, Gond lives were normalized within colonial mines and mining operations. On the other, they were simultaneously situated within the emerging deep history of the earth, in the depths of Gondwana. The visceral encounters with the earth in the coal mines of the nineteenth century gave new and naturalized connotations to Gond myths. The Legends of Lingo appeared in situ with the deep mining of Gondwana.

Forsyth's reading of Gond practices in the mines was distorted. As Gonds became laborers in the mines, their legends changed and acquired different motifs. The rise of modern colliery goddess cults or religious practices in the twentieth century was not just a reenactment of traditional tribal beliefs. It was a modern attempt to control the destiny of their precarious lives in the mining industry, an articulation of the risks involved, and a political critique of management of mines. Gonds have also maintained sustained protests against the development of mines in the region and their exploitative forms of management.[144]

Tribal myths have changed with changes in tribal social and political lives, further exposing Forsyth's romantic escapism. Laboring in the coal mines also led to the detribalization of the Gonds. Heterogeneous labor recruitment policies, economic changes, and even the transformation of the landscape due to the mines have all contributed to the process.[145] Gond cosmologies of creation themselves, which were co-opted within this European naturalized deep antiquity, were subsequently absorbed into their own cultural and political identity. Over the years, the Legends of Lingo became increasingly Hindu in character. This was partly because the tribes in many regions gave up their tribal language and adopted the local dialect and religious metaphors.

In the twentieth century, tribal anthropology itself moved away from its earlier urges to disconnect the tribes from Hindus and link them to the deep history of the land. Elwin, a mid-twentieth-century missionary turned anthropologist of the Gonds, denounced Hislop's and Chatterton's versions as "sheer nonsense," which he thought focused more on metaphysics than on social context.[146] For Elwin, the metaphysical references of the Legends of Lingo, which developed within the deep historical primitivism of the nineteenth century, had little relevance. He was more interested in understanding how the legends captured Gond society, rituals, and livelihood. His Lingo (collected from stories of the Bastar region) is, therefore, more a sociological entity who did not ascend to heaven but married seven wives and had children who populated the Bastar region. His story of Lingo is one of brotherly rivalries, which was a common theme in Gond society.[147]

Yet geological and mythological deep time and the concomitant questions of tribal habitat remain central to Gond identity and politics, particularly in the twenty-first century. Within the context of political and economic marginalization, their link with the land—and in turn with deep time—became the only legitimate theme in Gond political articulation of their identity to the state and to the public, to voice their conditions of deprivation and to protect their lives and livelihood. In the 1950s, the Gondwana movement began in different parts of central India in order to assert the political and cultural ambition of establishing a distinct Gond entity. These demands have come in response to the growing sense of loss of tribal land, culture, and rights within the modern Indian political economy and the dominant Hindu and Brahminical cultural assertions. These movements became more politically active in the 1990s, with clear demands for a separate Gondwana state for the Gonds.[148] This demand is complex because as a community Gonds are scattered widely across central India, without any clear political boundaries. What used to be known as Gondwana is now divided into various other states, inhabited by various communities. Because of the amorphous nature of their political and cultural aspiration for statehood, the search for a deep and imagined Gond homeland has appeared as the key feature of the Gondwana movement. Demands for a separate state for Adivasi (aboriginal) populations have become aligned with the deep sense of the Gond homeland, which in turn derives from the deep past of Gondwanaland. Mayuri Patankar has studied a wide range of contemporary Gond oral, visual, and literary traditions, in which the idea of "homeland" of the Gonds features prominently. These stories of creation no longer refer to Lingo or the formation of Gondwana but, deriving from Suess's and Wegener's ideas, stress the southern supercontinent of Gondwanaland as the original homeland of the Gonds. These recent invocations of the Gondwanaland are connected to the Gond revivalist movement around the religion of Punem, which seeks to establish a unified Adivasi religion and social system.[149] These narratives have all imbibed deep geological naturalism. A contemporary Gond poet writes, using Suess's and Wegener's themes of deep time and referring to the Tethys, the breakup of Pangaea, and Gondwanaland as the original southern homeland of the Gonds, distinct from that of the Aryans, to reclaim Gondwana for the Gonds:

I weave a story of this land, always at your feet is Onkar.
First I bow to Mother Earth, whose greatness our ancestors sang of.
That same tale I sing again and again. This land blesses me, so I sing.
The land is shaped as a sphere. Under the land is Pangea.
Three-fourth of the land is water. It is called Penthalsa [Panthalassa].
Over time, there were many upheavals. The land broke into different pieces.
In the north is the land called Angara,[150] the entire land in the south is
Gondwana.
Tethys Sea is thrown in between. It divides both the lands.
Inhabitants of Angara are called Arya, those of the Gondwana are Gonds.[151]

Conclusion

The planetary vision of Gondwanaland as a geological landmass was based on colonial experiences of the nineteenth century. It reflected the underlying division of the earth ushered in by European imperialism. In her book *Southern Theory*, Raewyn Connell critiques the northern intellectual traditions produced in "metropolitan" universities, which appear mainstream and yet fail to explain the social, cultural, and political experiences of the global south.[152] Against this, Connell identifies southern theories and southern intellectual identities shaped by their specific politics, history, and marginalization. Although her work cannot be reduced to the simple geopolitics of north-south, or even to that of Gondwanaland, she refers to the geological similarities of the southern continents, which according to her have shaped a common cultural inheritance among southern aboriginal/indigenous/subaltern cultures that were once all part of the supercontinent of Gondwanaland.[153] Witzel's frame of "Gondwana mythologies" similarly combines the geological metaphor of Gondwanaland with the aboriginal cultures of Australia, the Andaman Islands, Melanesia, and sub-Saharan Africa.[154] This cultural southernness is underpinned by deep encounters with nature in the nineteenth century.

Gondwana also highlights the essential synergy of the two dominant themes of nineteenth-century geology: the industrial search for minerals and the intellectual quest for deep time. Coal, as the imperial geologist Roderick Murchison wrote in 1866, was the "meter of power of modern nations."[155] Consequently, economic geology, particularly the colonial preoccupation with coal, has defined much of the early historiography of colonial geology in India and elsewhere. This literature suggests that geology in the colonies had predominantly economic implications. It demonstrates how the imperial search for coal and other minerals drove colonial geology.[156] At the same time, there have been attempts at writing a more holistic history of colonial geology, using the Humboldtian "encyclopedic" worldview. This literature emphasizes the environmental perspectives of colonial geology, which contained concepts of evolution, extinction, and general ecological perspectives.[157]

It is essential to reject this divide between economic and holistic geology and to locate the colonial history of coal, cotton, and diamonds as part of the intellectual pursuit of geological deep time and its ecological perspectives. Jason W. Moore, in his influential critique of the Anthropocene, has stressed the need to return to the problem of capital, or the Captilocene, as he puts it, to the origin of capitalistic exploitation of resources and labor, "the emergence of new relations of power, profit and re/production from the long sixteenth century."[158] Critical to this is the appearance of what Moore describes as "cheap nature," produced by the dual frugality in the human understanding of nature, in terms of its commercial price and ethical worth.[159] This chapter has highlighted the relationship between the capitalistic/colonial exploitation of nature and the emergence of deep nature in the nineteenth century. The "cheap nature" that Moore refers to evolved alongside, and often *because of*, the European discovery of deep nature. To give an example, the fossilized *Glossopteris* plant, which was the main indicator of prehistoric Gondwanaland, was also considered the primary contributor to the coal reserves of the Southern Hemisphere.[160] Lewis Leigh Fermor, an imperial geologist, spent his entire career in the first half of the twentieth century as a geologist of the GSI exploring the coal mines of Gondwana and conducting geological investigations in the Deccan traps. He also surveyed the mines in Kenya, South Africa, and Malaya. Based on this vast imperial experience, he believed that the key difference between Gondwanaland and Laurasia was the economic wealth that the former presented to the latter. Since in Gondwanaland the oldest rock formations were more exposed to the surface than those in "non-Gondwanaland," there was much greater availability of minerals in the former. He showed that five out of the seven most important sources of manganese, 99.9 percent of the world's annual output of diamond, and more than half of the world's gold came from Gondwanaland. All the richest gold mines were in Rand in South Africa, Kolar in India, and Morro Velho in Brazil.[161]

At the same time, these mines also appeared disenchanting to those very people who had found them, as they promised to disrupt the prehistoric landscape that surrounded them. Economic geology summoned a form of modernity that both created and went against the very idea of prehistoric Gondwana. Conceived at a time of growing detachment from a degenerative modernity, prehistory was a product of deep naturalism, a search for primitive nature, an escape into a simple form of life. It is possible to understand the deep naturalism behind the idea of the Anthropocene through this dual frame of enchantment with deep history and disenchantment with its concomitant economic quotient.[162]

William G. Atherstone was a doctor, botanist, and geologist based in the Cape Colony around the middle of the nineteenth century. He was also the first geologist to come across alluvial diamonds in the Karoo in the 1860s. In 1873, in the middle of the rush for diamonds and gold, as the sea of billowy mountains of the Karoo appeared to be studded with gems glistening glaringly in a yellow hue, he yearned for and found glimpses of life in its "primeval savage state." He saw that in the Karoo, the wild dog still roamed "untamed, unconquered, free" and was hunted and eaten by humans. Atherstone reflected that in this primitive and savage way of life there, one was yet to become the master of the other; both resided in the deep past, unaware of their future conjoined destinies in the civilized world.[163] As he looked deeper into the landscape, away from the surface glitter of gold, he could see the bones of giants, reptiles, tortoises, and crocodiles buried under the earth. They were also similar to those found in central India, "the same glossopteris . . . and to all appearance similar rock formations, and the same *coal* and *diamonds*!" He imagined the primitive Indo-African continent extending to the east across the oceans as his land of refuge. These fossils, which made the imagination of that grand continent possible, were, he believed, the true "nuggets of the gouph."[164]

Deep time became a mode of establishing the European self at the heart of colonial landscapes. European explorers sensed this primitivism in the seemingly changeless, vast, and overbearing landscapes of the real and imagined Gondwana. The geology and the aboriginal inhabitants of central India presented a primeval purity of life. They also generated mineral wealth and agricultural riches for the empire. These quests for primitivism were attempts to escape the rapidly transforming colonies, the environmental changes, and political movements and to return to a deep state of nature. The discovery of the deep time of Gondwana was therefore deeply contradictory and yet inherently colonial.

There are wider implications of this attempt at writing a denaturalized history of Gondwanaland, which are beyond the scope of this chapter and the book. It is possible to see similar confluences of geology and ethnology and of geological primitivism and tribal aboriginality in various parts of the imagined Gondwanaland, such as in South Africa, Australia, and in South

America.[165] Since Gondwanaland came into being and made sense as a comparative category, its full deconstruction would require a similarly global and comparative frame. This invites future investigations of how Laurasia and Gondwanaland reflected the north-south geopolitical divide between the colonizing and colonized regions of the late nineteenth-century colonial world. Such an approach would allow us to understand how these connections between land and people, imagined in these different sites, shaped ideas of primitivism and finally that of the southern continent. It would then be possible to write a new history of Gondwanaland as a southern cultural, ethnological, and colonial experience by using the real and imagined landscapes of Gondwanaland.

The New Deep History

This book has sought to be a denaturalized history of deep time. In the process, it has also highlighted the politics of deep history. At one level, the contours of a political history of the deep time of nature seem obvious. Nature has a political history. Historically, access to water and land has been a profoundly political issue, intensified by colonialism in certain parts of the world. The Anthropocene and climate emergency engender clear political intent: to put climate change at the heart of political agendas and academic discourses and to devise the new modes of activism that these challenges require.[1] Yet the written history of nature itself, more specifically of deep history, even when motivated by these agendas, has held itself outside these political questions of nature central to the very objects and lives whose deep pasts they trace. This is due to the assumption that deep history is beyond politics as much as it is beyond conventional history. It is *natural.*

Mainstream accounts of deep time have confirmed this naturalized narrative. They have traced the rise of geohistorical awareness and secular temporality among European savants from the eighteenth century. The story they tell has focused predominantly on the emergence of the geological notion of time. John McPhee coined the term "deep time" to define nature's own temporality, beyond historical imagination.[2] He drew inspiration from his observations of the Great Basin of Utah and Wyoming, a world seemingly untouched by humans and dense with geological features. There, nature is situated at a distance from humans and from politics and has seemingly remained so since deep time. This contraction of human history, what Tom Griffiths refers to as "the last inch of the cosmic mile,"[3] is the geologists' finest hour, a celebration of their understanding of the past, which has in turn shaped the modern understanding of history as a discipline. In his recent work, Rudwick has defined the emergence of deep time as the making of history itself as a science. The invention of deep time, he suggests, provided scientific validation or modification to existing historical chronology, periodization, and myth.[4] These themes have generated significant and insightful attempts at aligning deep history with history.[5] At the same time, some historians have pointed out that this deep history was in fact "all too human," a product of nineteenth-century industrialism and escapism.[6]

Parallel to this tradition, a distinct historical imagination of deep time has emerged around earth histories, human aboriginality, and environmental politics. This literature integrates human history with geological time, particularly that of the global south. I refer to this as the "new deep history," which has been the site of my main argument. I began this study as a conventional history of geology, to situate the emergence of deep historical thinking in India among discourses of geology, mythology, and philology. In the process I encountered the limits of a discipline that continues to be invested in European savants, museums, institutions, print cultures, and epistemologies. At the same time, this research found the entrenchment of the geological mode of thinking in the Indian subcontinent in its Hindu antiquarianism, sacred geographies, and tribal aboriginality. Therefore, I have used deep history not just as a mode of tracing the deepening of time but also as an opportunity to deepen the historical understanding of time. I have explored the role of the geological mode of thinking in the deep imagination of India—in its landscape, people, and antiquity. The stories narrated here are also simultaneously about rights over nature: who owns nature and the right to narrate its past. Each of the episodes discussed in this book—for example, the building of the colonial canal system, the Himalayan expeditions in search of commercial items and fossils, the search for minerals, cotton soil, and tribal labor in Gondwana—encompass the dual narratives of that nature, which is simultaneously commercialized and antiquated. I suggest that the two processes of the colonization and deepening of nature are symbiotic and mutually dependent.

The intellectual lineage I have traced, therefore, lies in the robust traditions of deep history in the global south, which are sustained by eclectic engagements with aboriginality, indigeneity, memory, tribal politics, folklore, and art.[7] The political narrative of deep history becomes particularly evident when these traditions are aligned with the histories of marginalization and displacement experienced during colonialism.[8] This deep history has emerged not only in disciplines such as genetics, geographic information system mapping, and paleoanthropology, but it has also integrated these disciplines with aboriginal and tribal politics, their historical and contemporary marginalization, and their loss of habitat and resources. This literature has alongside generated serious contemplation of how to accommodate deep history within historical imagination.[9] It has shown that deep history, as we know it, is a product of northern experiences of a linear and progressive deepening of time.

Against this intellectual tradition, this new deep history posits the immediacy and intimacy of the deep history of the south. Scholars have suggested that one of the keys to writing deep history, which is normally beyond historical tools and records, is to appreciate the Australian Aboriginal understanding of deep time. For many Aboriginal populations, the past is "something personal, familial, geological and omnipresent," unlike northern deep history, which appears distant and remote.[10]

There are several examples of this political immediacy of deep history in Africa, Australia, and South Asia, a number of which I have considered here.[11] Collectively, the scholars who have engaged with these examples have not only avoided the Eurocentrism of mainstream history of geology but have also conceived of deep history as a political discourse. Deep history is now a more pervasive and indispensable historical category. The experiences of the contemporary ecological devastation of major water systems, mountain ecologies, and forests, particularly in Asia, Africa, Australia, and South America, have led to a new awareness of the entrenched political exigencies that have shaped these deep ecological changes. This awareness was made possible by the use of the conventional notions of geological time to narrate ecological, human, and social histories.

To illustrate the political enactment of deep history, let us briefly refer to a recent development in India. In February 2019, the Supreme Court of India ordered the eviction of around a million tribal people from the forests and hills where they make their homes. This action derived from the law passed by the Indian parliament in 2006 (The Scheduled Tribes and Other Traditional Forest Dwellers Act, commonly known as the Forest Rights Act, or FRA), which gave legal rights over forestlands and their produce to tribal populations and forest-dwelling communities, provided they could prove that their families have lived there for at least three generations. The FRA started a complicated and fraught process of collecting information to substantiate the rights of these forest dwellers. The recent Supreme Court order is in response to the petitions submitted by wildlife activists, who argued that the act has led to gross misuse and illegal distribution of forestlands. They contended that the endangered and fragile wildlife and forest biodiversity of India was being further threatened by illegal squatting, particularly following the FRA of 2006. The Supreme Court decreed that many of the tribal populations, who also claim Adivasi or aboriginal status, are in fact encroachers on these forests and hills. Hence, it ordered their eviction.[12]

Both the FRA and the recent Supreme Court ruling have proved controversial.[13] Environmental activists and environmentalists have criticized the failure of implementation of the FRA, leading to land distribution to populations from outside these forestlands and the destruction of forest ecology. They have argued that these people are not the natural inhabitants of these forests.[14] Others have opposed the Supreme Court's ruling as being designed to further marginalize the tribal populations of India from their land and livelihood.[15] In the media and scholarly literature, such episodes are often portrayed as a conflict between the conservation of nature and the protection of tribal/human rights.[16] These conflicts in India are similar to the "green-black"[17] encounters in Australia between indigenous people and environmentalists. The relationships between the two are not straightforward, nor do these groups always share ideas about aims and objectives.[18] There also appears to be a global trend of framing the rights of aboriginal populations as in conflict with ecological conservation.[19] These debates are taking place in India and elsewhere alongside expansion of mining operations that have deep colonial roots in tribal and Aboriginal lands in India and Australia.[20] There is also an intellectual trend toward establishing Aryanism as the indigenous civilization of India and thereby incorporating Indian tribal aboriginality into it.[21]

The FRA episode indicates that some of the sites of new deep histories lie adjacent to those of the people, underlaid by the politics of the state. Most of these debates are addressed to the contemporary political context but also refer to the deep history of these communities and of these hills and forests.[22] This particular process of referring to the history of the earth to address political discourses, along with the urge to find recourse in the deep past, emerged with nineteenth-century colonialism. Colonialism introduced or redrew the lines in the soil between the settler and the aboriginal, the shallow and the deep, that have been a key component of the emergence of deep history in South Asia, Australia, and Africa.[23] In India, the forest lines, the so-called inner line, was first drawn by the colonial state in 1873. The line, which ran through various districts of northeastern India, was ostensibly drawn to protect tribal groups, to avoid conflicts between the tribal and nontribal populations. It was also to protect British economic interests in the forestlands. Inside that line, the purchase of land and access to natural resources such as rubber, tea, and other forest products were heavily restricted.[24]

As scholars have shown, the line drawn through forests, people, and resources was also drawn through time, between history and prehistory. The "inner line" was a civilizational one, designed to keep "'primitives' bound to their 'natural' space in the forests."[25] Communities living within and without the line were separated by a "time regime" between the modern and the primitive. Those living outside the line were subjects of the modern state, while those inside it lived within a deep colonial time, "where slavery, headhunting, and nomadism could be allowed to exist."[26]

These processes and their deeper historical trajectories cannot be traced through a naturalized "species" narrative. The conflict between protecting the ecology of the forest and protecting the homes of forest dwellers in South Asia and elsewhere shows that this simultaneous naturalization and denaturalization is an expression of politics, of power, and of colonial and postcolonial governance. Deep history, whether that of communities or of the land, can as much be bestowed as withheld.

Introduction

Epigraph. Grant, *The Gazetteer*, xii.

1. Falconer, "Abstract of a Discourse," 107. Such statements were not unique, particularly in an earlier era. The French naturalist Georges Cuvier believed fossils should be read like historical documents. See Rudwick, *Georges Cuvier*, 182–83. Cuvier, unlike Falconer, belonged largely to a period before the birth of prehistory, when antiquarianism and geology were much more closely aligned. Therefore, the historical analogy was more relevant in his time. In this book, we investigate the motives and implications of such shared antiquarian methods used to understand prehistory.

2. Rudwick, *Worlds before Adam*, 410–11. Donald Kelly has written about the long historical trajectory of the emergence of prehistory in "The Rise of Prehistory," 17–36.

3. The word "prehistory" appeared probably for the first time in English with the publication of Daniel Wilson's *Prehistoric Man*.

4. Montesquieu explained the historical evolution of laws by referring to environmental contingencies and "climate" (i.e., nonnaturals). De Secondat, *The Spirit of Laws*. On Bodin, see Blair, *The Theater of Nature*.

5. See the chapter by Anand Vivek Taneja, "Ruins and the Order of Nature" in his dissertation, "Nature, History, and the Sacred in the Medieval Ruins of Delhi." Also see Das, *Wonders of Nature*; Koch, "Netherlandish Naturalism," 29–37.

6. Fernand Braudel first used the term *géohistoire* in his 1941 article, "Géohistoire: La société, l'espace et le temps," reprinted in Braudel, De Ayala, and Braudel, *Écrits*, 68–114. The fuller exposition of the idea was in volume 1 of his book on the Mediterranean, *The Mediterranean and the Mediterranean World in the Age of Philip II*. For an analysis of the origins of Braudel's geohistory, see (in French) Ribeiro, "La genèse de la géohistoire chez Fernand Braudel"; Knight, "The Geohistory of Fernand Braudel."

7. Amrith, *Crossing the Bay of Bengal*, 12.

8. Beinart, *The Rise of Conservation in South Africa*, 39.

9. Singh, *Natural Premises*, 175.

10. For a detailed study of the various facets of the links between geography and history, see Ogborn, "The Relations between Geography and History."

11. Warde, Robin, and Sörlin, *The Environment*.

12. See Myllyntaus and Saikku, *Encountering the Past in Nature*.

13. Worster, *Nature's Economy*.

14. Worster, *Dust Bowl*.

15. Schama, *Landscape and Memory*, 37–74.

16. The term "naturalism" is used here in a different sense than that Roy Bhaskar used while suggesting the links between natural sciences and social sciences, or Bernard Lightman and others did to refer to Victorian scientific naturalism. Here it does not suggest the adherence to realism in understanding society or, in this case, the past. Nor is it to stress the secularism of Victorian "scientific naturalists," although elements of both are invariably present in this analysis. See Bhaskar, *The Possibility of Naturalism*; Dawson and Lightman, *Victorian Scientific Naturalism*. The term "naturalism" in this context has the same sense that we started the discussion with, that the deep history of nature became a foundation of historical imagination.

17. Introduction to "History Meets Biology," 1492.

18. Rudwick, *Bursting*.

19. The book discusses the relevant literature in detail in the respective chapters. Of the several works published in recent years, Andrew Shryock and Daniel Smail's *Deep History* is the

most comprehensive attempt at using the deep past as a global historical methodology. James C. Scott in his recent *Against the Grain* has identified the emergence of statehood in the deep history of human control over and monopoly of natural and human resources.

20. McGrath and Jebb, *Long History, Deep Time*.

21. See Robin, "Perceptions of Place and Deep Time in the Australian Desert"; Byrne, "Deep Nation"; Griffiths, "Travelling in Deep Time." Alison Bashford has shown how Australian historians are particularly adept at thinking in terms of deep time in "The Anthropocene Is Modern History."

22. Banerjee, *Politics of Time*.

23. Trautmann, "Indian Time, European Time."

24. Trautmann, *India*.

25. Rudwick, *Bursting*, 2.

26. Rudwick, *Bursting*, 5–8.

27. Cregan-Reid, "The Gilgamesh Controversy," 224–25.

28. Schama, *Landscape and Memory*, 560–70.

29. Oreskes, "The Scientific Consensus on Climate Change," 138; Dipesh Chakrabarty, "The Climate of History," 201.

30. Dipesh Chakrabarty, "The Climate of History," 201.

31. Byrne has shown that the modern Australian adoption of Aboriginal heritage, and in the process the nation's deep past, was part of a long-standing colonial project of appropriation of Aboriginal history ("Deep Nation").

32. Dipesh Chakrabarty, "Politics Unlimited," 242.

33. The Puranas are Hindu mythological texts dating from the fourth century, with historical notions of creation and destruction of the world.

34. Nayanjot Lahiri has made a similar point about Indian archaeology, which she refers to as "living antiquarianism." She has shown how various ancient sculptures continued to be displayed and worshipped in modern village temples and shrines in India, as well as in community practices, all of which became part of the European discovery of Indian antiquity in the nineteenth century ("Living Antiquarianism in India").

35. Some notable examples of that are Rudwick, *Worlds before Adam*; Cregan-Reid, *Discovering Gilgamesh*; and Charles Coulston Gillispie's classic book on the subject, *Genesis and Geology*.

36. Livingstone, *Adam's Ancestors*.

37. Piccardi and Masse, *Myth and Geology*.

38. Marshall, *The British Discovery of Hinduism*, 15–16.

39. Mamdani, *When Victims Become Killers*.

40. On the Ram Setu controversy, see Jaffrelot, "Hindu Nationalism." On the Saraswati, see Bhadra, Gupta, and Sharma, "Saraswati Nadi in Haryana."

41. Danino, "Genetics and the Aryan Debate"; Lal, "Aryan Invasion of India."

42. Ramaswamy, *The Lost Land of Lemuria*.

43. Nandy, "History's Forgotten Doubles."

44. Chakrabarti and Sen, "'The World Rests on the Back of a Tortoise.'"

45. Dipesh Chakrabarty, "Postcoloniality and the Artifice of History"; Merivirta-Chakrabarti, "Reclaiming India's History-Myth."

46. Gopal et al., "The Political Abuse of History."

47. Guha-Thakurta refers to archaeology as a "threatened science" ("Archaeology as Evidence," 3). Interestingly, she upholds archaeology as a "science," against myths. In fairness, she is referring to the various scientific dating methods used in archaeology. However, her approach also belies the belief that science is objective and incorruptible, unlike myths.

48. Guha-Thakurta, "Archaeology as Evidence," 10–11.

49. Bhattacharya, "Myth, History and the Politics of Ramjanmabhumi."

50. Bhattacharya, "Myth, History and the Politics of Ramjanmabhumi," 123.

51. Said, "Invention, Memory, and Place." Hans Bakker had seen the links between Jerusalem and Ayodhya ("Ayodhyā").

52. Rama's birthplace was identified at a certain site in Ayodhya in the nineteenth century. Bhattacharya, "Myth, History and the Politics," 136; Radhakrishna, "Of Apes and Ancestors," 8–9. Also see Mitra, *Prehistoric India*, 233–34.

53. See, for example, William Jones, "On the Gods of Greece, Italy and India."

54. Shryock and Smail, *Deep History*, 21–54, 160–90.

55. The Uniformitarian maxim of "The present is the key to the past," often ascribed to James Hutton, was based on the principle of uniformity and gradualism in change from the past to the present (O'Rourke, "A Comparison of James Hutton's Principles of Knowledge and Theory of the Earth"). I refer to a much more general phenomenon, which was common to geology, anthropology, and archaeology.

56. Manias, "The Problematic Construction of 'Palaeolithic Man.'"

57. Shryock and Smail, *Deep History*.

58. Dubow, "Earth History, Natural History, and Prehistory."

59. Keen, "The Anthropologist as Geologist," 79.

60. Keen, "The Anthropologist as Geologist," 83.

61. Edwin Bryant, *The Quest for the Origins of Vedic Culture*, chapter 7, "Linguistic palaeontology," 108–23.

62. Willey and Phillips, *Method and Theory in American Archaeology*, 2.

63. Showers, "A History of African Soil."

64. Asante, *Kemet, Afrocentricity, and Knowledge*.

65. Scott, *The Art of Not Being Governed*, xiv.

66. This is distinct from the argument of savaging the civilized, put forward by Ajay Skaria and Ramchandra Guha, which is premised more on colonial paternalism than on geographical determinism. Skaria, "Shades of Wildness"; Ramachandra Guha, "Savaging the Civilised."

67. Roberts, *From Massacres to Mining*.

68. For a study of the growing mineral industry and the dispossession of the tribal population in the postcolonial period, see Padel and Das, *Out of This Earth*. For a general study of marginalization and deprivation of tribal population in contemporary India, see Nandini Sundar, *The Burning Forest*.

69. Petraglia and Allchin, *The Evolution and History of Human Populations in South Asia*.

70. Banerjee, *Politics of Time*; Dasgupta, "'Heathen Aboriginals'"; Bates, "Race, Caste and Tribe in Central India"; Skaria, "Shades of Wildness"; and Damodaran, "Colonial Constructions."

71. Sumit Guha, "Lower Strata, Older Races, and Aboriginal Peoples."

72. Sumit Guha, "Lower Strata, Older Races, and Aboriginal Peoples"; see particularly 428.

73. Sumit Guha, "Lower Strata, Older Races, and Aboriginal Peoples," 433–38.

74. Manias, *Race, Science, and the Nation*, 5.

75. Rosenberg, "Toward an Ecology of Knowledge." For the integration of Rosenberg's disciplinary ecology with Latour's theory of network, see Akera, "Constructing a Representation."

76. Lenoir, "The Disciplines of Nature," 74.

77. For the quote, see Secord, "The Discovery of a Vocation," 133. A few examples of the predominantly nineteenth-century orientation of the history of geology are Greene, *Geology in the Nineteenth Century*; Secord, "King of Siluria"; Rudwick, *Bursting the Limits*; Secord, *Controversy in Victorian Geology*; Rupke, *The Great Chain of History*; Shortland, "Darkness Visible."

78. Srinivas and Panini, "The Development of Sociology."

79. Deepak Kumar, *Science and the Raj*, 31–112.

80. Hallam, "The Great Revolution," 410. Henry R. Frankel has studied at length this paradigmatic shift in earth sciences, *The Continental Drift Controversy*. David Sepkoski has

shown that throughout the nineteenth and early twentieth century, paleontology remained a predominantly "descriptive" evolutionary science. It was only in the 1970s that a group of American paleontologists defined a more distinctive discipline for themselves, known as "paleobiology," with greater theoretical scope and experimental parameters (*Rereading the Fossil Record*).

81. Trautmann, *Aryans and British India*, 131–64.

82. For such histories of prehistoric archaeology see, Dilip K. Chakrabarti, *India*; Sen and Ghosh, *Studies in Prehistory*; Dilip K. Chakrabarti, "Robert Bruce Foote and Indian Prehistory."

83. Rudwick, *Bursting*, 417–31, particularly 426–27.

84. See, for example, the two works Cunningham, *Inscriptions of Asoka*; and Horace Hayman Wilson, *Ariana Antiqua*. For a history of the formative years of Indian archaeology, see Guha-Thakurta, *Monuments, Objects, Histories*.

85. Guha-Thakurta, *Monuments, Objects, Histories*, 118–19.

86. De Terra, "Preliminary Report"; de Terra, "Pictorial History," 1.

87. Foote, *Collection*, iii.

Chapter 1 · The Canal of Zabita Khan

1. J. Colvin, "On the Restoration," 112.

2. "Papers regarding the reconstruction of the Canal of Zabeta Khan in the Ganges—Jumna," IOR/F/4/916/25799, 21–32, APAC, BL.

3. "Papers regarding the reconstruction of the Canal of Zabeta Khan," 16.

4. "Board of Commissioners at Farruckabad to the Honourable Vice President in Council of Fort William, 7 October 1809," in "Papers regarding the reconstruction of the Canal of Zabeta Khan," 35.

5. Lieutenant Tod, Surveyor Doab canal to Colonel Gastin (?), Surveyor General, 20 June 1810, in "Papers regarding the reconstruction of the Canal of Zabeta Khan," 55.

6. Lieutenant Tod, Surveyor Doab Canal to Colonel Gastin (?), Surveyor General, 20 June 1810, 23.

7. Lieutenant Tod, Surveyor Doab Canal to Colonel Gastin (?), Surveyor General, 20 June 1810, 57.

8. Sudeshna Guha has indicated the continuance of colonial traditions of archaeology and heritage in the postcolonial representations of antiquarianism in *Artefacts of History*. Similarly, Nayanjot Lahiri has explored how archaeology has formed an essential cultural identity of India in "Archaeology and Identity in Colonial India." Both Dipesh Chakrabarty and Ashis Nandy see Orientalist antiquarianism as the genesis of modern and Eurocentric historical tradition of India. See Dipesh Chakrabarty, "Postcoloniality and the Artifice of History"; and Nandy, "History's Forgotten Doubles."

9. For an introduction to the history and historiography of Indian antiquarianism of the eighteenth century and the emergence of Indian archaeology in the nineteenth century from these premises, see Sudeshna Guha, *Artefacts of History*, 1–66. Also see Indra Sengupta and Ali, *Knowledge Production*. Schnapp et al., *World Antiquarianism*, provides a comprehensive coverage of classical as well as prehistoric antiquarianism in China, Europe, India, Egypt, Mediterranean, Japan, and Polynesia.

10. Dilip K. Chakrabarti has written a comprehensive history of Indian archaeology in *India*. See also Sudeshna Guha, *Artefacts of History*. Other texts are referred to subsequently.

11. "Extract Bengal Revenue Consultations, 31 December 1822," in "Papers regarding the reconstruction of the Canal of Zabeta Khan."

12. Cautley, *Notes and Memoranda*, 5.

13. Herbert Michael Wilson, *Irrigation in India*, 47–48.

14. Tillotson, *The Artificial Empire*, 65.

15. Joyce Brown, "Smith, Robert."

16. Blunt, "A Description of the Cuttub Minar." The British had started preservation work on the Taj Mahal around the same time, see Etter, "Antiquarian Knowledge and Preservation."

17. Ewer, "An Account of the Inscriptions on the Cootub Minar," 481.

18. Ewer, "An Account of the Inscriptions on the Cootub Minar," 481.

19. Ewer, "An Account of the Inscriptions on the Cootub Minar," 485.

20. "Repairs made to the Kutb Minar and Jama Masjid at Delhi—suggestions of Major Robert Smith for the future preservation of the Kutb Minar—repair of a well attached to the Jama Masjid," IOR/F/4/1324/52472, March 1829–January 1831, APAC, BL.

21. Cole, *The Architecture of Ancient Delhi*, 81.

22. "The Jumna Canal near Meerut with soldiers and fortifications, 1808," National Army Museum, *Online Collection*, https://collection.nam.ac.uk/detail.php?acc=1971-08-16-1.

23. Such early colonial paintings have been generally interpreted as belonging to contemporary Romantic tradition of the "picturesque" (De Almeida and Gilpin, *Indian Renaissance*, 189–94). These do not refer specifically to Robert Smith's painting of the Yamuna Canal. This panting cannot be defined by the "picturesque" trope. It needs to be seen as a painting of a canal project by an engineer, and therefore its futuristic and modernistic (as opposed to the "picturesque") depictions are unmistakable.

24. Joyce Brown, "Smith, Robert," 637.

25. Cautley, *Notes and Memoranda*, 3–4.

26. Briggs, *History of the Rise of the Mahomedan Power*. For the passages on the canal and fossils, see 452–53.

27. Rennell, *Memoir*, 72–73.

28. Rennell, *Memoir*, 73–75.

29. Colvin, "On the Restoration," 106–27.

30. Colvin, "On the Restoration," 110.

31. Thornton, *Gazetteer* (1858), 467–68.

32. Hosagrahar, *Indigenous Modernities*, 94n.

33. Gupta, *Delhi between Two Empires*; Bayly, *Rulers, Townsmen and Bazaars*.

34. See Hosagrahar, *Indigenous Modernities*, 1–3, for these contrasting images of Delhi in the British imagination.

35. Quoted in Devji, "India in the Muslim Imagination," 13.

36. Gupta, "From Architecture to Archaeology," 58.

37. Naim, "Syed Ahmad and His Two Books."

38. Naim, "Syed Ahmad and His Two Books," 699.

39. For a history of the discovery of this antiquity, see Holt, *Thundering Zeus*, 67–80.

40. For details of Gerard and Burnes's travels, see "Continuation of the Route of Lieut. A. Burnes and Dr. Gerard, from Peshawar to Bokhara," *Journal of the Asiatic Society of Bengal* 2 (1833): 1–22.

41. Tod, "Indo-Graecian Antiquities," 13.

42. Tod, "An Account of Greek, Parthian, and Hindu Medals."

43. Joyce Brown, "A Memoir of Colonel Sir Proby Cautley," 195.

44. Cautley, *Notes and Memoranda*, 4.

45. Cautley, *Notes and Memoranda*, 4n.

46. Cautley, *Notes and Memoranda*, 12–14.

47. Muthahar Saqaf, "A Rock Solid Project"; Nirmal Sengupta, "Irrigation," 1920.

48. Stone, *Canal Irrigation in British India*, 14–15.

49. Arnold, *The Tropics*, 76, 108.

50. Stone, *Canal Irrigation in British India*; Gilmartin, "Scientific Empire and Imperial Science"; G. N. Rao, "Canal Irrigation."

51. Arnold, *The Tropics*, 75. In particular, see the chapter "Romanticism and Improvement," 74–109.

52. Bayly, *Empire and Information*, 44–96.

53. Grove, *Green Imperialism*, 347, 409–10.

54. Pratik Chakrabarti, *Western Science in Modern India*, 27–94.

55. *Khadirs* are the low-lying flood plains of the Doab region. The soil is formed mostly of fresh alluvial deposits from the river. *Bangars* are the higher lands lying beyond the flood plains. They consist of older alluvial deposits.

56. Cautley, *Report on the Central Doab Canal*, 1:3.

57. Cautley, *Report on the Ganges Canal Works*, 1:95. Also see Elizabeth Whitcombe, "Irrigation."

58. Cautley, "On the Use of Wells," 328.

59. Cautley, "On the Use of Wells," 328.

60. Cautley, *Report on the Ganges Canal Works*, 1:125–27.

61. Baird Smith, *A Short Account of the Ganges Canal*, 28–29.

62. Cautley, *Report on the Ganges Canal Works*, 1:288.

63. Barnett, "A Seal of Sri-vadra."

64. Atkinson, *Statistical, Descriptive and Historical Account*, 7.

65. Colvin, "On the Restoration," 106–27.

66. Colvin, "On the Restoration," 105–6.

67. Colvin, "On the Restoration," 106.

68. Colvin, "On the Restoration," 106–8.

69. Mackeson, "Journal of Captain C. M. Wade's voyage," 181.

70. Siddiqui, "Water Works and Irrigation System," 52.

71. Driver, "Yule, Sir Henry."

72. Yule, "A Canal Act."

73. Feroze Shah established the city of Forezepur on the banks of the Sutlej River in the fourteenth century.

74. Yule, "A Canal Act," 213–16.

75. Yule, "A Canal Act," 216.

76. The original texts he consulted were Dow, *The History of Hindostan*; Batuta, *Travels*.

77. Yule, "A Canal Act," 218.

78. Yule, "A Canal Act," 219.

79. Danino, *The Lost River*, 122–54. For a critique of these views, see Habib, "Imaging River Sarasvati."

80. Danino, The Lost River, 122–54; Habib, "Imaging River Sarasvati"; Vaidya, *Prehistoric River Saraswati*; Murthy, "The Vedic River Saraswati."

81. Chadha, "Conjuring a River, Imagining Civilisation," 62–63.

82. Briggs, *History of the Rise of the Mahomedan Power*, 175–77 and 450.

83. Briggs, *History of the Rise of the Mahomedan Power*, 175–77 and 450. For the passages on the canal and fossils, see 452–53.

84. *Descriptive sketch of the Sirhind Canal.*

85. Habib has referred to the "Sorrsuttee" mentioned in Persian maps: "Imaging River Sarasvati," 51.

86. Briggs, *History of the Rise of the Mahomedan Power*, 453.

87. C. F. Oldham, "The Saraswatī and the Lost River."

88. "Notes on the lost river."

89. "Notes on the lost river," 7–8.

90. "Notes on the lost river," 3–5.

91. "Notes on the lost river," 3.

92. "Notes on the lost river," 27.
93. "Notes on the lost river," 23–27.
94. Asif, *A Book of Conquest*, 31. On the minefield of names in this history, see Witzel, "Moving Targets?"
95. R. D. Oldham, "On Probable Changes," 340.
96. R. D. Oldham, "On Probable Changes," 323–24 and 326–27. Quotation on 326.
97. Martineau, *The Life and Correspondence*, 1:120.
98. R. D. Oldham, "On Probable Changes," 323; Cunningham, *The Ancient*, 253–83.
99. R. D. Oldham, "On Probable Changes," 323–24.
100. R. D. Oldham, "On Probable Changes," 327.
101. Asif, "The Long Thirteenth Century of the Chachnama," 485.
102. R. D. Oldham, "On Probable Changes," 331–32.
103. R. D. Oldham, "On Probable Changes," 329–32. Rajputana literarily meant "the land of the Rajputs," who are a Hindu military caste of western India.
104. R. D. Oldham, "On Probable Changes," 340.
105. R. D. Oldham, "On Probable Changes," 341–42.
106. R. D. Oldham, "On Probable Changes," 341.
107. C. F. Oldham, "The Saraswatī and the Lost River," 55–62, 76.
108. Müller, *Sacred Books of the East*, 32:58–60. See particularly 60.
109. Müller, *Sacred Books of the East*, 32:58–60. See particularly 60.
110. C. F. Oldham, "The Saraswatī and the Lost River," 51–52.
111. C. F. Oldham, "The Saraswatī and the Lost River," 52.
112. C. F. Oldham, "The Saraswatī and the Lost River," 52.
113. C. F. Oldham, "The Saraswatī and the Lost River," 61.
114. C. F. Oldham, "The Saraswatī and the Lost River," 76.
115. Danino, *The Lost River*, 268.
116. Bhadra, Gupta, and Sharma, "Saraswati Nadi in Haryana."
117. "Haryana Govt Pumps 100 Cusec Water."
118. Pilgrim, "Suggestions concerning the History."
119. Wadia, "The Tertiary Geosyncline," 93.
120. Wadia, *Geology of India for Students*, 5–6.
121. Vaidya, *Prehistoric River Saraswati.*
122. Cautley, "Further Account," 222.
123. Cautley, "Discovery of an Ancient town."
124. Cautley, "Further Account," 222–23.
125. Cautley, "Further Account," 224–6.
126. Cautley, "Discovery of an Ancient town," 43.
127. "Indo-Scythians" referred to a race that was believed to have lived in Central and South Asia from the second century BC to the fourth century AD. Orientalist scholars discovered epigraphic and numismatic evidence of them in India in the nineteenth century. (Cunningham, "Coins of the Indo-Scythians"). Tod found philological and ethnological evidence of Scythian races in Rajasthan (*Annals and Antiquities of Rajasthan*, 1:15 and 277).
128. Prinsep, "Bactrian and Indo-Scythic Coins"; Prinsep, "On the Ancient Roman Coins."
129. Prinsep, "Note on the Coins."
130. Prinsep, "On the Connection," 624.
131. Cautley, "Discovery of an Ancient town," 43–44.
132. "Cautley to J. Thomaston, Secretary to the Governor General, 7 May 1838," in "Government of India forward to the British Museum in London a large collection of fossils and geological specimens made by Captain Proby Thomas Cautley, Superintendent of the Doab Canal," IOR/F/4/1866/79252, APAC, BL.

133. "Cautley to J. Thomaston." Cautley quoted these paragraphs from Briggs's *History of the Rise of the Mahomedan Power*, 1:452–53, which had, in turn, gathered this information from Farishta's book. The same account is also available in Dow, *The History of Hindostan*, 1:343.

134. "Proceedings of the Asiatic Society," *Journal of the Asiatic Society of Bengal* 3 (1834): 529.

135. Rudwick, *Worlds before Adam*, 228–29, 408–9.

136. Tournal used the French term *antéhistorique*, whose idiomatic translation was "prehistory." See Chippindale, "The Invention of Words."

137. Rudwick, *Worlds before Adam*, 408–10.

138. Cautley, *Notes and Memoranda*, 177–78.

139. "The Jumna Canals," 57.

140. E. Smith, "Notes on the Specimens," 623–26.

141. E. Smith, "Notes on the Specimens," 624.

142. E. Smith, "Notes on the Specimens," 630.

143. "Proceedings of the Asiatic Society," 529.

144. Dean, "On the Strata."

145. Dean, "On the Fossil Bones."

146. Dean, "On the Fossil Bones," 496–97.

147. *Gazetteer of the Delhi District*, 21.

148. Dean, "On the Fossil Bones," 500.

149. Spry, *Modern India*, 1:340.

150. Spry, *Modern India*, 1:338–41.

151. He was skeptical about whether some of these belonged to humans, see Prinsep, "Note on the Proceedings," 502.

152. Prinsep, "Occurrence of the Bones," 632–33.

153. Prinsep, "Occurrence of the Bones," 635. Quotation on 633.

154. Prinsep, "Occurrence of the Bones," 635.

155. Lahiri, *Decline and Fall*. See in particular the section "Environment and Collapse," 139–263.

156. Giosan et al, "Fluvial Landscapes."

157. Bhadra, Gupta, and Sharma, "Saraswati Nadi in Haryana," 273–88.

Chapter 2 • Ancient Alluviums

1. Burkill, "Chapters," 869–70. Also referred to in Richard H. Grove, *Green Imperialism*, 412.

2. "Arrangements regarding the administration of the Botanic Garden at Saharanpur-Dr George Govan is appointed Superintendent at a salary of 200 rupees 'per mensem', etc," IOR/F/4/587/14218, APAC, BL.

3. Baird Smith, *Canals of Irrigation*, 10.

4. Braudel, *The Mediterranean*, 1:25–52.

5. D. N. Wadia wrote, "All the great Himalayan rivers are older than the mountains they traverse" (*Geology of India for Students*, 19).

6. Murchison, *Palæontological Memoirs*, 1:xxv–xxvi.

7. Woodward and Harrison, "Royle, John Forbes."

8. Nathaniel Wallich to John Adam, Secretary, Medical Board, April 21, 1826, forwarding a letter from Royle to Wallich dated April 7, 1826, Board's Collection, 1827–8, IOR/F/4/955, 124, APAC, BL.

9. Wallich to Lushington, May 1, 1826, Board's Collection, 129.

10. Royle, *Essay*, 26.

11. Royle, *Illustrations*, 1:vii.

12. Royle, *Illustrations*, 1:xxix.

13. Murchison, *Palæontological Memoirs*, 1:xxvii.

14. Murchison, *Palæontological Memoirs*, 1:xxvii.

15. Cautley and Falconer, "Notice on the Remains of a Fossil Monkey," 502.

16. "Extracts from Dr. Royle's Explanatory Address on the Exhibition of his collections in Natural History, at the Meeting of Asiatic Society, on the 7th March," *Journal of the Asiatic Society of Bengal* 1 (1832): 97.

17. Grove, *Green Imperialism*, 347, 409–10.

18. Kalapura, "Constructing the Idea of Tibet." Also see Bishop, *The Myth of Shangri-La*.

19. Samuel Turner, *An Account of an Embassy*.

20. Webb, "Memoir," 294.

21. Moorcroft, "A Journey to Lake Mánasaróvara."

22. There is a growing history of science literature on the Himalayas. Lachlan Fleetwood has stressed the "fragility" of the early nineteenth-century Himalayan expeditions, as British explorers struggled with the remoteness and the high altitudes of the mountains as well as their deteriorating tools and failing health, in " 'No Former Travellers.' " Thomas Simpson has highlighted the "fraught" and "uncertain" nature of imperial mapmaking in the remote Himalayas and the northeastern hilly regions of India, in " 'Clean Out of the Map.' " As this and the next chapter will show, the natural history of the Himalayas identified through such imperial frailties nonetheless transformed ideas about the geohistory of the mountains, the subcontinent, its human evolution, and the future imagination of the Indian nation.

23. "Miscellaneous, Original and Select: Proceedings of Societies," *Asiatic Journal and Monthly Register for British and Foreign India, China and Australasia* 7 (1832): 148–49.

24. Lloyd and Gerard, *Narrative*, 2:122.

25. Falconer, "Official report," 1:557–76, 573–74.

26. Falconer, "Official report," 558.

27. Cautley, "On the Structure," 268.

28. See, for example, the issue of *Science in Context* 22 (2009) on science in the mountains. In particular, see the introduction to the issue by Charlotte Bigg, David Aubin, and Philipp Felsch.

29. Greene, *Geology in the Nineteenth Century*, 144–60.

30. Bigg, Aubin, and Felsch, "Introduction." There is recent scholarship on science and the Himalayas, referred to in note 22. For a comprehensive historiographical analysis of some of this scholarship, see Simpson, "Modern Mountains."

31. Herbert, "On the Organic Remains," 266.

32. Hodgson and Herbert, "Account."

33. Herbert, "On the Organic Remains," 266.

34. Falconer and Cautley, "Sivatherium Gigantium."

35. Murchison, *Palæontological Memoirs*, 1:xxviii–xxix.

36. Cautley, "On the Fossil Remains."

37. Sandes, *The Military Engineer in India*, 2:275–77.

38. "Proceedings of the Asiatic Society," *Journal of the Asiatic Society of Bengal* 3 (1834): 590–94, 592–94.

39. Crawfurd, *Journal*, 935–37.

40. "Extract from a Paper written for the Medical and Physical Society of Bombay to accompany specimens forwarded to James Prinsep Esquire Secretary, Asiatic Society," May 18, 1836, Bombay Proceedings, IOR/P/347/46, no. 37, APAC, BL.

41. Hügel and Fulljames, "Recent Discovery," 288–90.

42. Charles Lush to James Prinsep, April 22, 1836, IOR P/347/46 no. 36, APAC, BL.

43. Fulljames, "Note on the discovery."

44. Hügel and Fulljames, "Recent Discovery," 290–91.

45. "Dr Nicholson on the Island of Perim," *Journal of the Bombay Branch of the Royal Asiatic Society* 1, no. 1 (1841): 11.

46. Baird Smith, "On the Structure," 342.
47. Baird Smith, "On the Structure," 339.
48. "Government of India forward to the British Museum in London a large collection of fossils and geological specimens made by Captain Proby Thomas Cautley, Superintendent of the Doab Canal," IOR/F/4/1866/79252, 1, APAC, BL.
49. "Letter from Hugh Falconer submitting a report to East India Company Directors on his work on the Siwalik collection of fossils in the British Museum," Mss Eur F447, f9, India Office Records and Private Papers, British Library, APAC, BL.
50. Cautley and Falconer, *Fauna Antiqua Sivalensis.*
51. Falconer, "Description of some Fossil Remains," 484.
52. Geologists in Europe used the term "tertiary" from 1815 to describe a "former world" to which they assigned the growing discovery of fossils. See Rudwick, *Bursting*, 543–54. It was also used in the early nineteenth century to refer to mountains of more recent formation.
53. Falconer, "Abstract of a Discourse," 109.
54. Falconer to Charles Lyell, November 29, 1856, Lyell 1, Box: GEN 110, 1294, Papers of Charles Lyell, GB 0237 Sir Charles Lyell, Edinburgh University Library Special Collections Division.
55. Lyell, *Elements of Geology*, 133–34; Rupke, *The Great Chain of History*, 31–41.
56. Rudwick, *Bursting*, 600–621.
57. Cautley, *Ganges Canal*, iii.
58. Cautley, *Ganges Canal*, 12–14. Quotation on 20.
59. Cautley, *Ganges Canal*, 14.
60. Baker and Durand, "Sub-Himalayan Fossil."
61. Baker and Durand, "Sub-Himalayan Fossil," 741.
62. Cautley and Falconer, "On Additional fossil species," 358.
63. Kennedy and Ciochon, "A Canine Tooth."
64. I will discuss these in detail in chapter 4, in the section "Humans as Apes."
65. Falconer and Cautley, "Notice on the Remains of a Fossil Monkey," 500–502.
66. Falconer and Cautley, "Notice on the Remains of a Fossil Monkey," 503.
67. Murchison, *Palæontological Memoirs*, 1:xxx.
68. Murchison, *Palæontological Memoirs*, 1:24.
69. Falconer, "Abstract of a Discourse," 108.
70. Tickell, "List of Birds," 582–83.
71. Falconer and Cautley, "Notice on the Remains of a Fossil Monkey," 502.
72. Rudwick, *Georges Cuvier*, 260.
73. William Jones, "Discourse," ix–xvi, xii–xiii.
74. For a study of the scientific pursuits of the Asiatic Society, see Pratik Chakraborty, "Asiatic Society and Its Vision of Science."
75. Murchison, *Palæontological Memoirs*, 1:1.
76. Rudwick, *Bursting*, 417–31.
77. Murchison, *Palæontological Memoirs*, 1:2.
78. Falconer, "Abstract of a Discourse," 107.
79. Murchison, *Palæontological Memoirs*, 1:1–2. The full text of the lecture is on 1–29.
80. Murchison, *Palæontological Memoirs*, 1:1–2.
81. Almond, "Druids, Patriarchs, and the Primordial Religion," 392.
82. Almond, "Druids, Patriarchs, and the Primordial Religion," especially 380.
83. Higgins, *The Celtic Druids*, 281.
84. Wilford, "An Essay on the Sacred Isles" (1811), 121–29.
85. Burrow, "A Proof that the Hindus had the Binomial Theorem." Simon Schaffer explains these European deliberations on Asiatic sciences and this Indo-centrism in British Orientalism

as attempts to establish Newtonian philosophy as *the* universal and primordial cosmology. He explains, "An important use of these sciences was to show how the new coloniser was in some sense already present in the colonised territory." Schaffer, "The Asiatic Enlightenments of British Astronomy," 51–52, 77.

86. Maurice, *Indian Antiquities*, vol. 6, part 1, 242–47, particularly 245.
87. Falconer, *Descriptive Catalogue*.
88. Falconer, *Descriptive Catalogue*, 2–5. Quotation in 4.
89. Lydekker, *Catalogue of the Remains*, 1:i–ii.
90. Lydekker, *Catalogue of the Remains*, 1:i–ii.
91. Lydekker, *Catalogue of the Remains*, 1:i–iii.
92. Lydekker, *Catalogue of the Remains*, i, iii, iv.
93. Falconer, "Primeval Man," 2:585.
94. Falconer, *Descriptive Catalogue*, 7.
95. Murchison, *Palæontological Memoirs*, 2:643.
96. Leopold, "British Applications."
97. Leopold, "British Applications," 581–82.
98. Trautmann, *Aryans*, 174.
99. Müller, "On the relation of the Bengali to the Arian," 319–50.
100. Falconer, "On the asserted Occurrence," 381.
101. Falconer, "On the asserted Occurrence," 382.
102. Falconer, "On the asserted Occurrence," 381.
103. Falconer, "On the asserted Occurrence," 382.
104. Falconer, "On the asserted Occurrence," 382.
105. Falconer, "Primeval Man," 570–600. The editor in his note explains that Falconer wrote this in 1863 as an introduction to a work on the remote human antiquity, which was never published (570). Falconer, "On the asserted Occurrence."
106. Falconer, "Primeval Man," 576.
107. Falconer, "On the asserted Occurrence," 388.
108. Falconer, "Primeval Man," 578. Emphasis added.
109. Falconer, "Primeval Man," 578.
110. Falconer, "Primeval Man," 579.
111. Falconer, "Primeval Man," 579.
112. Falconer, "On the asserted Occurrence," 382.
113. Leonard G. Wilson, "Brixham Cave."
114. Falconer, "Report of Progress in the Brixham Cave."
115. Kapila, "Race Matters," 479.
116. Trautmann, *Aryans*, 167.
117. Kapila, "Race Matters."
118. Prestwich, *Geology*, 2:534–35.
119. William Jones, "On the Gods of Greece, Italy and India," 271.
120. "Extract of a letter from J. Nimmo, Esq., Bombay, dated Colaba, 29th August, 1846," *Calcutta Journal of Natural History* 7 (1847): 359.
121. Arthur Bedford Orlebar, "Some Observations," 239.
122. Horace Hayman Wilson, "Lecture," 194.
123. W. T. Blanford, "The African element."
124. W. T. Blanford, "The African element," 293–94.
125. W. T. Blanford, *Observations*.
126. Henry F. Blanford, "On the age and correlation."
127. Sclater, "The Mammals of Madagascar."
128. Ramaswamy, *The Lost Land of Lemuria*, 38.

129. Suess, *The Face of the Earth*, 1:595–97.
130. Lydekker, "Notices of Siwalik Mammals," in particular 66–70.
131. Lydekker, "Further notices of Siwalik Mammalia," 40.
132. Lydekker, "Further notices of Siwalik Mammalia," 40.
133. Lydekker, "Further notices of Siwalik Mammalia," 40–41.
134. Suess, "Are Great Ocean Depths Permanent?," 183.
135. Wadia, *Geology of India for Students*, 203–4.
136. Woodward, "The Antiquity of Man," 213; Barrell, "Probable Relations of Climatic Change."
137. *Yale North India Expedition*, 8:3–9.
138. De Terra, "Stone Age Tools found in India."
139. Lewis, "Preliminary Notice of New Man-Like Apes from India"; Gregory, Hellman, and Lewis, *Fossil Anthropoids*.
140. Kennedy, *God-Apes and Fossil Men*, 93–120.
141. Pilgrim, "Preliminary Note on a Revised Classification."
142. Pilgrim, "Correlation of Siwaliks."
143. Pilgrim, "New Siwalik Primates," especially 3.
144. Sahni, "The Himalayan Uplift," 59–61. Quotation on 57. In the nineteenth century, George William Traill had come across *silajit* alongside fossil bones and shaligrams in central Himalayas ("Statistical Sketch of Kamaon," 159).
145. Sahni, "The Himalayan Uplift," 57
146. Napier and Weiner, "Olduvai Gorge and Human Origins."

Chapter 3 · Mythic Pasts and Naturalized Histories

1. Rossi, *The Dark Abyss of Time*, 8–9. For an analysis of the strong influence of Mosaic philosophy on seventeenth-century science in general, see Blair, "Mosaic Physics." Also see Marshall, *The British Discovery of Hinduism*, 15–16.
2. There is a vast literature on the influence of myths in European geology. Rudwick's several books; Livingstone, "The Preadamite Theory"; Gillispie, *Genesis and Geology*. More recently, Vybarr Cregan-Reid has highlighted Suess's reference to the deluge in the nineteenth century and the afterlife of the myth of Gilgamesh, discovered by the Assyriologist George Smith, in Victorian art and literature. Cregan-Reid, *Discovering Gilgamesh*, 195–201. However, he does not elaborate on how and whether Deluge myths influenced Victorian geology.
3. See the sources in the previous note. Also Smail, "In the Grip of Sacred History."
4. Ernst, "India as a Sacred Islamic Land." Faisal Devji has shown that India remained a critical theme in Islamic cartographic and visual presentation even in the nineteenth century ("India in the Muslim Imagination," 13).
5. Cook, *Matters of Exchange*.
6. Jardine, *Ingenious Pursuits*, 103.
7. Livingstone and Withers, *Geography and Enlightenment*.
8. Nasr, *Religion and the Order of Nature*, 130–32.
9. Foucault, *Order of Things*, 141.
10. Latour, *We Have Never Been Modern*, 30.
11. Of several works on the relationship between science and religion in Europe, some of the key ones are Frank Turner's classic study of the Victorian conflict between scientific naturalism and religious ideas, *Between Science and Religion*, and Bernard V. Lightman's examination of the complexity of agnosticism in "Huxley and Scientific Agnosticism. More recently, Jason Ā. Josephson-Storm has reinstated the significant role of myths and occult ideas at the birth of philosophy, anthropology, sociology, folklore, psychoanalysis, and religious studies in *Myth of Disenchantment*.

12. Said, "Invention, Memory, and Place."
13. Jazeel, *Sacred Modernity*.
14. Feldhaus, *Connected Places*; Eck, *India*; Ara Wilson, "The Sacred Geography of Bangkok's Markets."
15. For a more detailed analysis see Pratik Chakrabarti, "Purifying the River."
16. Cannon, *The Collected Works of Sir William Jones*, 2:3–4.
17. Cannon, *The Life and Mind of Oriental Jones*, 80.
18. Marshall, *British Discovery of Hinduism*, 15–16.
19. Marshall, *British Discovery of Hinduism*, 35.
20. App, *The Birth of Orientalism*, xii.
21. Said, *Orientalism*.
22. William Jones, "On the Gods of Greece, Italy and India," 271.
23. William Jones, "On the Gods of Greece, Italy and India," 222.
24. William Jones, "On the Gods of Greece, Italy and India," 236.
25. William Jones, "On the Gods of Greece, Italy and India," 221.
26. Kapil Raj, "British Orientalism in the Early Nineteenth Century, or Globalism versus Universalism," in *Relocating Modern Science*, 139–58.
27. William Jones, "On the Gods of Greece, Italy and India," 198.
28. Daniel Bassuk suggests that these deliberations started in the seventeenth century in the works of Baldaeus but refers to Jones's as one of the early attempts in modern times to link avatarism and incarnation formally (*Incarnation*, 42–47, 104–7). Several scholars have analyzed the similarities and dissimilarities between Hindu avatarism and Christian incarnation—how both represented immanent yet transcendent and earthly connotations of the descent of the divine. For example, see Sheth, "Hindu Avatāra and Christian Incarnation: A Comparison"; and Parrinder, *Avatar and Incarnation*.
29. Jacob Bryant, *A New System*, 3:587–88.
30. William Jones, "On the Origin," 484.
31. William Jones, "On the Origin," 484.
32. William Jones, "On the Gods of Greece, Italy and India," 234.
33. William Jones, "On the Gods of Greece, Italy and India," 214.
34. William Jones, "The Third Anniversary Discourse."
35. William Jones "On the Chronology," 127.
36. William Jones "On the Origin."
37. William Jones "On the Origin," 484.
38. William Jones "On the Origin," 484–85.
39. Panigrahy, Panigrahy, Dash, and Padhy, "Ethnobiological Analysis from Myth to Science."
40. Rudwick, *Bursting*, 333–36, 445–70; Gillespie, *Genesis and Geology*.
41. Moor, *Hindu Pantheon*, 182.
42. Maurice, *The History of Hindostan*, vol. 1; Coleman, *The Mythologies of the Hindus*, chapters 2–4.
43. Maurice, *The History of Hindostan*, 1:561.
44. Maurice, *The History of Hindostan*, 1:contents page.
45. Trautmann, *Aryans*, 76–80.
46. Trautmann, *Aryans*, 80.
47. Picart, *Ceremonies*, following 414.
48. Baldaeus, *A True and Exact*, 849–50.
49. Bassuk, *Incarnation*, 105–6.
50. Bassuk, *Incarnation*, 105–6.
51. Stolte, *Philip Angel's Deex-Autaers*, vol. 5.

52. Sanjay Subrahmanyam has provided a detailed analysis of Picart's engravings with European landscape in "Monsieur Picart and the Gentiles of India," 197–214.
53. Baldaeus, *A True and Exact*, 849–50.
54. Patterson, "Enlightenment, Empire and Deism."
55. App, *The Birth of Orientalism*, 440–80.
56. William Jones to Sir John Macpherson, Chittagong, 6 May 1786, in Cannon, *The Letters of Sir William Jones*, 2:698.
57. *The Works of Sir William Jones*, 13:322.
58. *The Works of Sir William Jones*, 13:324.
59. *The Works of Sir William Jones*, 13:245.
60. *The Works of Sir William Jones*, 13:325.
61. Said, "Invention, Memory, and Place."
62. Mabbett, "The Symbolism of Mount Meru," 66.
63. Jones, "On the Chronology," 260–61.
64. Wilford, "On Egypt," 295.
65. Leask, "Francis Wilford and the Colonial Construction of Hindu Geography," 206–14.
66. "Remarks on the Preceding Essay by the President," 467.
67. Wilford, "On Mount Caucasus."
68. Wilford, "An Essay on the Sacred Isles," (1808).
69. The phrase is borrowed from Wujastyk, "'A Pious Fraud.'"
70. See for example, Tilak, *The Arctic Home in the Vedas*.
71. Thapar, "The Theory of Aryan Race and India."
72. Ramakanta Chakrabarty, *Vaisnavism in Bengal*, 396.
73. Javed Majeed, *Ungoverned Imaginings*.
74. Turner, *An Account of an Embassy*.
75. Colebrooke, "introductory note," 375.
76. Colebrooke, "introductory note," 376.
77. Colebrooke, "On the Heights of the Himálaya Mountains."
78. Arnold, *The Tropics and the Traveling Gaze*, 3–73.
79. Arnold, *The Tropics and the Traveling Gaze*, 67–70.
80. Arnold, *The Tropics and the Traveling Gaze*, 69–70.
81. Arnold, *The Tropics and the Traveling Gaze*, 73.
82. Ward, *A View of the History*, 2:6–7.
83. Shaffer, *"Kubla Khan" and "The Fall of Jerusalem,"* 46.
84. Ramaswamy, *The Lost Land of Lemuria*, 137–81.
85. Wilford, "On Mount Caucasus."
86. Pavgee, *The Vedic Fathers*.
87. Pavgee, *The Vedic Fathers*, 34–38, 63–68.
88. Pavgee, *The Vedic Fathers*, 25–30. This dating of the Vedas is inaccurate. The earliest Vedas have been dated to 1200 BC or between 1500 and 500 BC. See Flood, *An Introduction to Hinduism*, 37–38; and Witzel, "Vedas and Upaniṣads," 68.
89. Pavgee, *The Vedic Fathers*, 25–30.
90. Roy, *Prithibir Puratattva* [The facts of the ancient earth], vol. 2. Roy also wrote at length on the Ayurvedic conceptions of the human body, the humors, and the environment. See Mukharji, *Doctoring Traditions*, 134–35.
91. Mabbett, "The Symbolism of Mount Meru."
92. Sleeman, *Rambles and Recollections*, 1:157.
93. Maurice, *Indian Antiquities*, 5:908.
94. Maurice, *Indian Antiquities*, 5:908.

95. Maurice, *Indian Antiquities*, 5:909. Although Maurice refers to the "man-lion" as "Rama-Avatars," the seventh avatar, the former is usually associated with Narasimha, the fourth avatar.

96. Maurice, *Indian Antiquities*, 5:909.

97. Wilford, "On the ancient Geography of India," 413.

98. Moor, *Hindu Pantheon*, 309–10.

99. Moor, *Oriental Fragments*, 85–86.

100. Oakley, "Folklore of Fossils."

101. Pymm, " 'Serpent Stones.' "

102. Rudwick, *Bursting*, 624.

103. Rudwick, *Bursting*, 65, 243.

104. Knell, *The Culture of English Geology*, 305–6.

105. William Smith, *Strata Identified by Organised Fossils*, 1.

106. Secord, *Controversy in Victorian Geology*, 312–18; Rudwick, *The Great Devonian Controversy*, 5.

107. Garzanti, "Stratigraphy and Sedimentary History."

108. *Narrative of a Journey from Caunpoor to the Boorendo Pass*, 2:122.

109. Colebrooke, "On the Valley of the Setlej River," 358, 379–80.

110. Traill, "Statistical Report on the Bhotia Mehals of Kamaon," 15.

111. Herbert, "On the Organic Remains," 271–72.

112. "List of Donors and Donations to the Museum of the Asiatick Society," *Asiatick Researches* 12 (1816): xx–xxviii, xxi–xxii, xviii.

113. "List of the Donors and Donations to the Museum of the Asiatick Society, (from January, 1820)," *Asiatick Researches* 14 (1822): n.p.

114. "Fossil Ammonites from India, called by the Hindoos Salagram," donated by Dr Wallich, on June 4, 1822, in "A List of Donations to the Library, to the Collections of Maps, Plans, Sections and Models; and to the Cabinet of Minerals," *Transactions of the Geological Society of London*, 1 (1824): 425–39, 437.

115. Cowper Reed, "Sedgwick Museum Notes," 256–57.

116. Cowper Reed, "Sedgwick Museum Notes," 257.

117. Cowper Reed, "Sedgwick Museum Notes," 256–61.

118. Colebrooke, "On the Heights of the Himálaya Mountains."

119. Colebrooke, "On the Heights of the Himálaya Mountains," 264.

120. Colebrooke, "On the Religious Ceremonies of the Hindus," 240.

121. "A List of Donations to the Library, to the Collections of Maps, Plans, Sections and Models; and to the Cabinet of Minerals, belonging to the Geological Society," *Transactions of the Geological Society of London* 5 (1821): 625–50, 643.

122. "Donations to the Royal Asiatic Society of Great Britain and Ireland, from its Institution, March 15, 1823, to March 15, 1827," *Transactions of the Royal Asiatic Society of Great Britain and Ireland*, 1 (1826): 600–40, 638–39.

123. Balfour, *Cyclopædia*, 73.

124. For the rise of Avataric evolutionism in India, see C. M. Brown, "Colonial and Postcolonial Elaborations of Avataric Evolutionism." For the rise of Vaishnavism in the nineteenth century, see Ramakanta Chakrabarty, *Vaisnavism in Bengal*, 403–9.

125. Ramakanta Chakrabarty, *Vaisnavism in Bengal*, 396

126. C. M. Brown, "Colonial and Post-colonial," 720–26.

127. Several scholars have commented on the compatibility established in the nineteenth century between Darwinian evolution and Hindu reincarnation. David L. Gosling shows that Hindu scholars such as Vivekananda and Aurobindo were influenced by Darwinian ideas and

incorporated them within Puranic reincarnation. Gosling, "Science and the Hindu Tradition?"; C. M. Brown, *Hindu Perspectives on Evolution*; Edelmann, *Hindu Theology and Biology*.

128. Van der Geer, Dermitzakis, and de Vos, "Fossil Folklore from India."

129. Ramachandra Rao, *Śālagrāma-kosha*, 22–25.

130. Chandrasekharam, "Geo-mythology of India." Chandrasekharam comments on the similarities between the ammonite fossil and Vishnu's *chakra*: "The ammonite fossil with circular shape and radiating ribs look very similar to Vishnu chakra" (31).

131. Dorothy Vitaliano, "Geomythology," 5.

132. Mayor, *The First Fossil Hunters*, 15–53.

133. Mayor, *The First Fossil Hunters*, 129–35, especially 135.

134. Van der Geer, Dermitzakis, and de Vos, "Fossil Folklores from India," 75.

135. Chandrasekharam, "Geo-mythology of India," 29.

136. Panigrahy, Panigrahy, Dash, and Padhy, "Ethnobiological Analysis from Myth to Science," 181.

137. Peter Young, *Tortoise*, 46.

138. Tylor, *Researches into the Early History*, 331–36.

139. Tylor, *Researches into the Early History*, 333.

140. Rappenglück, "The Whole World Put between Two Shells."

141. Peter Young, *Tortoise*, 47–51.

142. Locke, *An Essay concerning Human Understanding*, 1:242–43.

143. Davidson, "White Versions of Indian Myths and Legends."

144. Peter Young, *Tortoise*, 31–37.

145. Bird, *Bones for Barnum Brown*, 118.

146. Gould, *The Structure of Evolutionary Theory*, 1–33, 745–50. For a critique of Gould's suggestion, see Chakrabarti and Sen, " 'The World Rests on the Back of a Tortoise,' " 827.

147. Murchison, *Palæontological Memoirs*, 1:xxvii.

148. Murchison, *Palæontological Memoirs*, 1:372–74.

149. *Proceedings of the Zoological Society of London* 2 (1844): 55.

150. Falconer and Cautley, "Notice on the Remains of a Fossil Monkey."

151. Falconer and Cautley, "Notice on the Remains of a Fossil Monkey," 504. Falconer's assertion of the pterodactyl as a mythological animal came before Richard Owen's references to it as the "flying reptile, or dragon" of Greek myths, in his *Geology and Inhabitants of the Ancient World* (11–12).

152. Murchison, *Palæontological Memoirs*, 1:388. Emphasis in original.

153. Murchison, *Palæontological Memoirs*, 54–55, 83–88.

154. Horace Hayman Wilson, *The Vishnu Purana*.

155. For references to Hindu evolutionary thinking and avatars, see Pavgee, *Vedic Fathers*, 11, 100.

156. *The London, Edinburgh, and Dublin Philosophical Magazine*, 532.

157. *Proceedings of the Zoological Society of London*, 12 (1844): 55.

158. *Proceedings of the Zoological Society of London*, 12 (1844): 85–86.

159. *Proceedings of the Zoological Society of London*, 12 (1844): 87.

160. *Proceedings of the Zoological Society of London*, 12 (1844): 87.

161. *Proceedings of the Zoological Society of London*, 12 (1844): 87.

162. *Proceedings of the Zoological Society of London*, 12 (1844): 86.

163. McGowan-Hartmann, "Shadow of the Dragon."

164. Owen, *Geology and Inhabitants*, 11–12. As far as it is evident from the sources, Owen's lectures at the Zoological Society did not contain these references to mythological creatures, or he certainly did not elaborate on them in his scholarly publications to the extent that Falconer did.

165. Murchison, *Palæontological Memoirs*, 1:1.
166. Falconer, "Abstract of a Discourse," 108.
167. Falconer, "Abstract of a Discourse," 110–11.
168. Lyell, *Principles of Geology*, 1–8.
169. Lyell, *Principles of Geology*, 8.
170. Rudwick, *Bursting*, 394–95.
171. Murchison, *Palæontological Memoirs*, 2:586.
172. Tylor, *Researches into the Early History*, 306–32.
173. Pavgee, *Vedic Fathers*, 10–14. Quotation on 14.
174. Pavgee, *Vedic Fathers*, 18.
175. Lyell, *A Manual of Elementary Geology*, 183–84.
176. A particular fragment is at the Indian Museum in Kolkata. "Colossochelys," *Geological Survey of India*, accessed November 7, 2019, http://museum.gsi.gov.in/cs/VirtualMuseum/articles/1463315739569?resolvetemplatefordevice=true. The locality there is shown to be "Piram Island, Gujarat." There is also an entire shell on display in the upper floor of the museum. Parts of the shell are also in Punjab or in Himachal Pradesh. Badam, "Colossochelys Atlas," 149.
177. See, for example, Lydekker, "On the Land Tortoises of the Siwaliks," 209.
178. Barnum Brown, "The Largest Known Land Tortoise," 183.
179. Barnum Brown, "The Largest Known Land Tortoise," 184–85.
180. Barnum Brown, "The Largest Known Land Tortoise," 185.
181. Barnum Brown, "The Largest Known Land Tortoise," 187.
182. Sachau, *Alberuni's India*, 1:248.
183. Scafi, *Mapping Paradise*, 342–64. As Scafi shows, from the nineteenth century, the mapping of paradise became more of a cartographical than theological preoccupation.

Chapter 4 · Remnants of the Race

1. Rivett-Carnac, "On Stone Implements," 221–22.
2. Rivett-Carnac, "On Stone Implements," 222.
3. Rivett-Carnac, "On Stone Implements," 222.
4. The term "bushman" was derived from the Afrikaans word *boschjesman* (meaning "forest-man") and was usually used by early European settlers to refer to the indigenous populations in South Africa, particularly the San people.
5. Robert B. Young, *The Life and Work of George William Stow*, 27–28.
6. Goodrum, "The History of Human Origins Research," 341.
7. Lane Fox, *Catalogue of the Anthropological Collection*, xiii–xiv.
8. Boule, *Fossil Men*, 361. Emphasis added.
9. Browne, "Missionaries and the Human Mind."
10. Bowler, "From 'Savage' to 'Primitive.'"
11. *Celts* were the stone or metal implements shaped like chisels or ax heads, usually associated with prehistoric humans.
12. Edward B. Tylor's conclusions about Stone Age cultures of Australia, New Zealand, and the Americas were based not on his own anthropological observations of these people or of the geological features of these regions, but on that of others (*Researches into the Early History*, 192–228). He did visit Mexico early in his career and wrote the *Anahuac: Or, Mexico and the Mexicans, Ancient and Modern*. *Anahuac* is essentially a travelogue and does not contain discussions on Stone Age cultures or prehistory. Similarly, Lubbock did not actually confront any of the "modern savages" of the Pacific Islands or Greenland that he wrote about within his evolutionary scheme of racial development (*Prehistoric Times*, 427–577).
13. Bashford, "Deep Genetics," 319.
14. Skaria, *Hybrid Histories*, 1–35.

15. Tom Griffiths, "Travelling in Deep Time."

16. There is a vast literature on aboriginality and colonialism, of which this is only a small sample: Rose, *Dingo Makes Us Human*; Ghosh, "A Market for Aboriginality"; Mawani, "Genealogies of the Land"; Head, *Second Nature*.

17. Karlsson and Subba, *Indigeneity in India*; Baviskar, *In the Belly of the River*; Sundar, *The Burning Forest*.

18. Banerjee, *Politics of Time*; Dasgupta, "'Heathen Aboriginals'"; Bates, "Race, Caste and Tribe."

19. Kennedy, *God-Apes and Fossil Men*; Dilip K. Chakrabarti, *India*, chapters 2–3.

20. Pels, "The Rise and Fall of the Indian Aborigines."

21. Sumit Guha, "Lower Strata, Older Races, and Aboriginal Peoples," 433.

22. Sumit Guha, "Lower Strata, Older Races, and Aboriginal Peoples," 433–38, especially 436 and 438. See also Kennedy, *God-Apes and Fossil Men*, 358–80.

23. Lane Fox, *Catalogue of the Anthropological Collection*, xiii–xiv.

24. Gould, "The Geometer of Race." In the expanded edition of *The Mismeasure of Man*, Gould referred to this as "racial geometry" (401–12). Also see Pratik Chakrabarti, *Medicine and Empire*, chapter 4, for an overview of the historical constitution of ideas of race.

25. While this chapter focuses on central India, the argument, as will be evident, can apply to several other parts of the world, such as Australia, South Africa, or South America, wherever questions of aboriginality were juxtaposed to those of prehistory.

26. Ball, *Jungle Life in India*, 9.

27. Lake, *Sir Donald McLeod*, 68.

28. Lake, *Sir Donald McLeod*, 46–47.

29. Lake, *Sir Donald McLeod*, 58.

30. Lake, *Sir Donald McLeod*, 62.

31. Hislop Manuscripts, Hislop Mss 8958, 49, 54–55.

32. Hislop Manuscripts, Hislop Mss 8958, 49, 54–55.

33. Dasgupta, "'Heathen Aboriginals.'"

34. Trautmann, *Aryans*, 206–7.

35. Dravidianism has been a key theme in historical discussions on human antiquity in central and southern India. Crispin Bates has shown that in the early nineteenth century Dravidian racial theories were a core part of European ethnological deliberations on early humans in the subcontinent ("Race, Caste and Tribe"). Trautmann has shown how Dravidianism played a critical role in the anthropological work of Lewis Henry Morgan. European missionaries incorporated tribes within and it signified tribal aboriginality ("Dravidian Kinship"). Ramaswamy, however, has shown that in the early twentieth-century Indian (Tamil) deliberations, Dravidianism became closely associated with Tamil identity and the idea of their ancestral land (*Passions of the Tongue*, 62–77).

36. Caldwell, *A Comparative Grammar*, v, 23, 38, 69–73.

37. Oppert, *On the Original Inhabitants*, vii.

38. Oppert, *On the Original Inhabitants*, 9.

39. Oppert, *On the Original Inhabitants*, iv, 94, 109.

40. Briggs, "Two Lectures."

41. Trautmann, *Aryans*, 160.

42. Dasgupta, "'Heathen Aboriginals.'"

43. On the Orientalist discovery of Hinduism, see Marshall, introduction to *The British Discovery of Hinduism*, 35–38, 40–41.

44. Arnold, *The Tropics and the Traveling Gaze*, 69–70.

45. Hislop, "A Model Missionary Address, to the students of the New College, Edinburgh," December 3, 1859, Hislop Manuscripts, Mss 8958, 155.

46. Hislop, "A Model Missionary Address," 156–57.

47. Temple, *Papers related to the Aboriginal Tribes*, part 1, 4.
48. Kennet, "Observations on the Language of the Gonds," 33.
49. Kennet, "Observations on the Language of the Gonds," 33.
50. Driberg and Harrison, *Narrative of a Second visit to the Gonds.*
51. Kennet, "Observations on the Language of the Gonds," 43.
52. Hislop, "Essay on the Hill Tribes."
53. Hislop, "Essay on the Hill Tribes," 27.
54. Trautmann, *Aryans*, 155–60. Stevenson was more sympathetic of the Vaishnavic and Hindu traditions. See Philip Constable, "Scottish Missionaries."
55. Trautmann, *Aryans*, 174.
56. Bose, "Chhattisgar," 272.
57. Elwin, *The Aboriginals.*
58. "The Late Mr. H. W. Voysey."
59. William King, "Notice of Pre-Historic Burial Place," 179.
60. William King, "Notice of Pre-Historic Burial Place," 180.
61. George Smith, *Stephen Hislop*, 242.
62. George Smith, *Stephen Hislop*, 242.
63. George Smith, *Stephen Hislop*, 242–43.
64. Meadows Taylor, *Descriptions of Cairns*, 329.
65. Meadows Taylor, *Descriptions of Cairns*, 337.
66. Meadows Taylor, *Descriptions of Cairns*, 339. See also 342–43, 347–48.
67. Meadows Taylor, *Descriptions of Cairns*, 342–43.
68. Meadows Taylor, *Descriptions of Cairns*, 353 and 356.
69. Manias, *Race, Science*, 137–68.
70. Wright, *The Celt*. See in particular the images on 52, 54, 55, 57, and 64.
71. Meadows Taylor, *Descriptions of Cairns*, 347–48. The same paper was published in the *Transactions of the Royal Irish Academy* 24 (1873): 329–62.
72. Rivett-Carnac, "Prehistoric Remains in Central India," 1.
73. Rivett-Carnac, "Prehistoric Remains in Central India," 6–8.
74. Meadows Taylor, "Ancient Remains," 183.
75. Wright, *The Celt*, 5, 45–51.
76. For the disappearance of the notion of common origin, see Manias, *Race, Science*, 169–201.
77. William King, "Notice of Pre-Historic Burial Place," 183–84.
78. William King, "Notice of Pre-Historic Burial Place," 182–84, especially 184.
79. Ball, "Stone Monuments," 291.
80. Dalton, "Rude Stone Monuments," 112.
81. Dalton, "Rude Stone Monuments," 114.
82. Dalton, "Rude Stone Monuments," 114–15. It is not clear whether Peppé himself took the photograph, which Dalton subsequently sent to the Asiatic Society.
83. *Proceedings of the Asiatic Society of Bengal: January to December, 1873* (Calcutta, Baptist Mission Press, 1873), 131.
84. *Proceedings of the Asiatic Society of Bengal*, 132–33.
85. Preface to Temple, *Papers Relating to the Aboriginal Tribes*, vii.
86. Dalton, "Rude Stone Monuments," 130–34.
87. Unnamed letter, Chiculdah Western Super Mare to Hislop, November 23, 1859, Hislop Manuscripts, MSS 8957, 476–7.
88. Kennet, "Observations on the Language of the Gonds," 52.
89. Huxley, "On the Geographical Distribution," 404–5. Lubbock confirmed this view in the second edition of his *Prehistoric Times* (London: Hertford, 1869), 378.

90. Trautmann, "Dravidian Kinship."

91. Dixon, *The Racial History of Man*. The idea of Indians belonging to proto-Australoids became increasingly popular in the 1930s among Indian intellectuals. See Chatterji, "Contributions from different Language-Culture Groups," 1:76–92.

92. Biraja Sankar Guha, "An Outline of the Racial Ethnology of India." Also see Biraja Sankar Guha, "Racial Affinities of the People of India."

93. *Census of India*, 10–12. Also, see Chatterji, "The Internationalism of India"; Chatterji, "India and Polynesia."

94. *Constituent Assembly Debates*. I am grateful to Dr. Sanjukta Das Gupta for this information.

95. Witzel, *The Origins of the World's Mythologies*, 216–18.

96. Sonakia and Biswas, "Antiquity of the Narmada *Homo Erectus*."

97. Sankhyan et al., "New Human Fossils and Associated Findings."

98. Armitage et al., "The Southern Route."

99. Lane Fox, *Catalogue of the Anthropological Collection*, xiii–xiv.

100. Qureshi, *Peoples on Parade*, 185–221.

101. "Proceedings of the Asiatic Society of Bengal for September 1866," *Proceedings of the Asiatic Society of Bengal*, 187–89, 188.

102. "Proceedings of the Asiatic Society of Bengal for September 1866," 187–89, 188.

103. "Proceedings of the Asiatic Society of Bengal for September 1866," 188–89.

104. Bates, "Race, Caste and Tribe," 19.

105. Lane Fox, *Catalogue of the Anthropological Collection*, xiii–xiv. Or as the British zoologist and museum curator, William Henry Flower described isolated aboriginal populations of the islands of the Indian Ocean as "living fossils" who could be used as analytical tools for understanding human antiquity: "We may, however, look upon the Andamanese, the Aetas, and the Semangs, as living fossils, and by their aid conjecture the condition of the whole population of the land in ancient times (*Essays on Museums*, 304). Also quoted in David Tomas, "The Production of Ethnographic Observations," 75–76. I am grateful to Cam Sharp Jones for pointing out this reference to me.

106. Lubbock, *Prehistoric Times*, 427–50.

107. See for example, the special issue, guest edited by Bruce Buchan and Linda Andersson Burnett, "Knowing Savagery," in *History of the Human Sciences*.

108. Darwin, *Descent of Man*, 1:3, 174–75.

109. Darwin, *Descent of Man*, 1:9–33.

110. Falconer, *Descriptive Catalogue*, 7; Falconer, "Primeval Man," 2:572, 585.

111. Falconer, "Primeval Man," 2:578.

112. Lyell, *The Geological Evidences*, 79.

113. Lyell, *The Geological Evidences*, 476–79.

114. Lyell, *The Geological Evidences*, 78–79.

115. Lyell, *The Geological Evidences*, 89.

116. Peterson, "Studying Man and Man's Nature."

117. Primate societies have remained useful for understanding prehistoric or Paleolithic human societies (Gamble, "Palaeolithic Society").

118. Sahlins, "The Social Life of Monkeys," 55.

119. Sahlins, "The Origin of Society," 77.

120. Sahlins, "The Origin of Society," 77.

121. "The Late Mr. H. W. Voysey."

122. Piddington, "Memorandum," 209–10.

123. Piddington, "Memorandum," 209–10.

124. Piddington, "Memorandum," 210.

125. Radhakrishna, "Of Apes and Ancestors," 8–9. Also see Mitra, *Prehistoric India*, 233–34.

126. Tylor, *Primitive Culture*, 1:343–44.

127. Radhakrishna, "Of Apes and Ancestors," 9–10.

128. For an analysis of the human-animal continuum observed by the British in India, see Pratik Chakrabarti, "Beasts of Burden."

129. *Report of the Ethnological Committee*, 3.

130. Ball provided a detailed account of several wolf-boys he came across in different parts of central India. Ball, *Jungle Life in India*, 458-60.

131. For a study of the search for the so-called missing link among South African aboriginal tribes, see Dubow, *Scientific Racism*, 20–65.

132. Viswanath, *The Pariah Problem*, xi. Caldwell believed that the *pariahs* were one of the oldest races in the subcontinent, the earliest of the Dravidian population, who were subsequently enslaved by successive waves of settlers. Their present "servile" status, according to him, indicated their deep prehistory in the subcontinent (*A Comparative Grammar*, 491–502).

133. Yule and Burnell, *Hobson-Jobson*, accessed July 24, 2019, http://dsalsrv02.uchicago.edu/cgi-bin/philologic/contextualize.pl?p.1.hobson.1764242.

134. Ball, *Jungle Life in India*, 676–77.

135. Ball, *Jungle Life in India*, 427–28.

136. Ball, *Jungle Life in India*, 677.

137. As mentioned by Lyell, *The Geological Evidences*, 94.

138. Gamble and Moutsiou, "The Time Revolution"; Kennedy, *God-Apes and Fossil Men*, 19; Leonard G. Wilson, "Brixham Cave."

139. Boule, *Fossil Men*, xi.

140. Cohen, "Charles Lyell," 88.

141. Nilsson, *The Primitive Inhabitants of Scandinavia*. Burnett, "Sami Indigeneity." I am grateful to Andersson Burnett for this information.

142. Bulstrode, "The Industrial Archaeology of Deep Time."

143. Grant, *The Gazetteer of the Central Provinces*, xliv–xlv.

144. Grant, *The Gazetteer of the Central Provinces*, xlv.

145. Grant, *The Gazetteer of the Central Provinces*, xlv.

146. Rivett-Carnac, "Flint Implements from Central India," 27.

147. Lyell, *The Geological Evidences*, chapter 4, 59–74. Also see Lyell, "On the Occurrence."

148. For Evans's unconventional approach to dating European stone implements, see Bulstrode, "The Industrial Archaeology of Deep Time."

149. "Remarks on two flints from Jubbulpore," 199.

150. "Remarks on two flints from Jubbulpore," 199.

151. "Remarks on two flints from Jubbulpore," 200.

152. *Proceedings of the Asiatic Society of Bengal*, 144.

153. Rupert Jones, "Miocene Man in India," 349.

154. Foote, *On the Occurrence*, 1–4.

155. Foote, *On the Occurrence*, 10.

156. "Review: *On the Occurrence*."

157. *Proceedings of the Asiatic Society of Bengal*, 139–40.

158. *Proceedings of the Asiatic Society of Bengal*, 141.

159. Sen and Ghosh, *Studies in Prehistory*; Dilip K. Chakrabarti, "Robert Bruce Foote and Indian Prehistory." Chakrabarti refers to a few preceding works by Ball, Meadows Taylor, and Congreve, but he does not discuss them in any detail.

160. Foote, *On the Occurrence*, 26–27.

161. Foote, *Collection*, 2:2, 8

162. Foote, "On the Distribution."

163. Foote, "On the Distribution," 494.

164. Thomas Huxley was there probably drawing from his own excursions into the South Pacific two decades earlier.

165. Foote, "On the Distribution," 494.

166. Foote, *Collection*, 2:113

167. Foote, *Collection*, 2:1.

168. Foote, *Collection*, 2:60.

169. Foote, *Collection*, 2:185–86.

170. Foote, *Collection*, 2:40–42. Here Foote drew mainly from Thomas Holditch's *Gates of India*, 140–42.

171. Grafton Elliot Smith, *The Ancient Egyptians*, 81.

172. Foote, *Collection*, 2:185–86.

173. Robert B. Young, *The Life and Work of George William Stow.*

174. Stow, *The Native Races of South Africa*, 24.

175. Stow, *The Native Races of South Africa*, 24–25.

176. Stow, *The Native Races of South Africa*, 25.

177. Newton, "On the Study of Archaeology," 1.

178. As Meira Gold has recently shown, the German theologian Charles Josias von Bunsen continued to contest British geologist Leonard Horner's geological assertions of human antiquity in the 1860s ("Ancient Egypt," 218–21).

179. For example, Thangaraj, Macaulay, and their teams conducted DNA research among tribal populations in the Andaman Islands and the Malayan peninsula, respectively, to identify the routes of the earliest human migration out of Africa, see, K. Thangaraj et al., "Reconstructing the origin of Andaman Islanders"; Macaulay et al., "Single, Rapid Coastal Settlement."

180. Bergen, "What Arc You?"

181. Shryock and Smail, *Deep History*, 34.

Chapter 5 · The Other Side of Tethys

1. Suess, *Face of the Earth*, 1:596.

2. Suess, "Are Great Ocean Depths Permanent?"

3. Such references appeared throughout Alexander de Humboldt's *Personal Narrative*, vol. 4. See for example, 2, 23–30, 158, 225–70, 421, 511.

4. Secord, *Controversy in Victorian Geology*, 299–310.

5. Neubauer, "Gondwana-Land Goes Europe"; Leviton and Aldrich, "Contributions of the Geological Survey of India"; Monroe and Wicande, *The Changing Earth*; McLoughlin, "*Glossopteris*"; Frankel, *The Continental Drift Controversy*; Greene, *Alfred Wegener.*

6. Suess, *Face of the Earth*; du Toit, *Our Wandering Continents*; Wegener, *The Origin of Continents.*

7. On the debates around the continental drift theory, see Oreskes, *The Rejection of Continental Drift.*

8. Suess, *Face of the Earth*, 1:595–97. It is useful to point out here that "Indo-Africa" acquired other connotations in the twentieth century, in a genre of literature, particularly by Cyril Hromnik, that highlighted long historical trading connections between India and Africa. See Hall and Borland. "The Indian Connection."

9. Suess, *Face of the Earth*, 1:595–97.

10. Suess, *Face of the Earth*, 1:595–97.

11. Suess, *Face of the Earth*, 596.

12. Suess, "Farewell Lecture," 272.

13. Kious and Tilling, *This Dynamic Earth*, 3–5.

14. Wegener, *The Origin of Continents*, 192. The original German version, *Die Entstehung der Kontinente und Ozeane*, was published in 1915 and translated into English in 1924. In the

original German version, Wegener had used the term *Urkontinent*, which meant the original or a paleo supercontinent. "Pangæa," used in the English translation was a similar paleontological category, derived from a Greek term meaning "all-earth" or "mother-earth."

15. Withers, "Geography, Enlightenment, and the Paradise Question," 81.

16. Rupke, "Paradise and the Notion of a World Centre."

17. Suess referred to the biblical deluge and various indigenous traditions of it in his *Face of the Earth* (2:17–50). Vybarr Cregan-Reid has located in Suess's work one of the last attempts, at the end of the nineteenth century, of providing a geological reading of the deluge, *Discovering Gilgamesh*, 198–200).

18. Suess, *Face of the Earth*, 2:1906, part 3, 1.

19. Dubow, *A Commonwealth of Knowledge*, 208.

20. Vanderpoorten et al., "The Ghosts of Gondwana."

21. Witzel, *The Origins of the World's Mythologies*, 279–356.

22. This is not to ignore the histories of aboriginality and displacement experienced by people in the global north, such as the Native Americans, the Inuit, and the Sami people, which are as significant as those of the global south. The point being made here is that Gondwana captures the particular histories of such expressions of northern colonialism across the southern continents. It is possible to use Gondwana, since it is already a metonymical category, not just as a geographical expression but also as an antipodal one, within which the experiences of northern aboriginality can also be related as antithetical to northern mainstream histories.

23. *The Ain i Akbari by Abul Fazl Allami*, 2:222–23, 309. For a precolonial history of Gondwana and the Mughal and Maratha political settlements with Gonds. See Archana Prasad, "Military Conflict and Forests."

24. Dalton, *Descriptive Ethnology of Bengal*, 275.

25. Medlicott and Blanford, *A Manual of the Geology of India*, 1:97.

26. Cyril Fox, "The Gondwana System," 1, 4–5.

27. Blunt, "Narrative of a Route," 96.

28. Blunt, "Narrative of a Route," 111.

29. Thornton, *Gazetteer*, 2:355.

30. Grant, *The Gazetteer of the Central Provinces*, xiii.

31. Grant, *The Gazetteer of the Central Provinces*, xi.

32. Grant, *The Gazetteer of the Central Provinces*, xii. Grant added in footnotes, adopting from Richard Jenkins's *Report on the Territories of the Rajah of Nagpore* (referred to later), that the Bandarwas go entirely naked and "are said to destroy their relations when too old to move about, and eat their flesh, when a great entertainment takes place, to which all the family is invited" (34). Others slightly higher up the tribal ladder ate only victims of alien tribes. He left it open as to whether they actually practiced cannibalism, although he said that he himself did not find any trace of it.

33. In using the phrase "cultural landscape," we are reminded of Schama's *Landscape and Memory*.

34. Sleeman, *Rambles and Recollections*, 1:6, 131.

35. Lake, *Sir Donald McLeod*, 33–34.

36. Lake, *Sir Donald McLeod*, 10.

37. Lake, *Sir Donald McLeod*, 30–34.

38. Henry F. Blanford, "On the age and correlation." The name "Vindhyans" was given by Thomas Oldham in "Proceedings of the Asiatic Society of Bengal," *Journal of the Asiatic Society of Bengal* 25 (1856): 251.

39. Hislop Manuscripts, Mss 8958, 199. This is not the most widely accepted explanation. There are several alternative theories about the naming of Gonds.

40. Richard Jenkins to Elphinstone, August 27, 1811. Nagpur, Letter books of Sir Richard Jenkins (1785–1853), Mss Eur E111–12; F34, APAC, BL, 153

41. "Military operations conducted against the Gond tribes of the Narbada Territories by the Nagpur Subsidiary Force—final settlements made with the Gond chiefs," September 1818–June 1823, vol. 1, IOR/F/4/755/20541, APAC, BL.

42. Lt. Colonel to J. W. Adams, Commanding Nerbuddah field force to Lt Col Nicoll, Adjutant General of the Army, February 10, 1819, in "Military operations conducted against the Gond tribes," 1:77–78.

43. Jenkins to Hastings, Governor General, September 3, 1819, in "Military operations conducted against the Gond tribes," 1:137.

44. Jenkins to Hastings, 141–42.

45. Jenkins to Hastings, 135–36.

46. Jenkins to Hastings, 150.

47. Brigadier General Sir J. Malcolm to Major William Henley, January 22, 1820, in "Military operations conducted against the Gond tribes," vol. 2, IOR/F/4/755/20542, 177–78.

48. The *malguzari* system was the land revenue system adopted by the British in the Central Provinces from the 1860s. It was a form of revenue arrangement with the *malguzar*, or tenants, adopted from the revenue system of the Marathas, their predecessors.

49. See, Bhukya, "Enclosing Land."

50. Captain Montgomerie to Jenkins, July 16, 1819, in "Military operations conducted against the Gond tribes," 1:169. Also "Letter from Jenkins to Hastings, 3 September 1819," in "Military operations conducted against the Gond tribes," 1:135–36.

51. Captain Montgomerie to Jenkins, 171–95.

52. Brigadier General Sir John Malcolm to Major William Henley, January 22, 1820, in "Military operations conducted against the Gond tribes," 2:171–73.

53. Malcolm, *A Memoir of Central India*, 1:31–32. Nandini Sundar has made the important point that the region that she studies, Bastar, is culturally and linguistically diverse (*Subalterns and Sovereigns*, 1–20).

54. Jenkins, *Report on the Territories of the Rajah of Nagpore*, 29–30.

55. Preface to Temple, *Papers related to the Aboriginal Tribes*, iii.

56. "Proceedings of the Asiatic Society," *Journal of Asiatic Society of Bengal* 25 (1856): 249–52.

57. Forsyth, *The Highlands of Central India*, 141.

58. Lake, *Sir Donald McLeod*, 30–31.

59. Letter books of Sir Richard Jenkins (1785–1853), Bombay Civil Service 1800–28, Resident at Nagpur 1807–27, Mss Eur E111–12; F34: 1807–1817, Jenkins to Elphinstone, August 27, 1811, Nagpur, 152–55, 154, APAC, BL.

60. Sumit Guha, "Lower Strata, Older Races, and Aboriginal Peoples," 423–24.

61. Grant, *The Gazetteer of the Central Provinces*, xliii–xliv. Medlicott and Blanford, *A Manual of the Geology of India*, 1:371.

62. George Smith, *Stephen Hislop*, 215. Emphasis added.

63. Damodaran, "The Politics of Marginality"; Ramachandra Guha, "The Prehistory of Community Forestry"; Béteille, "Caste and Citizen."

64. Béteille, "Caste and Citizen."

65. Oppert, *On the Original Inhabitants*, v.

66. Sumit Guha, *Environment and Ethnicity in India*.

67. Sumit Guha, *Environment and Ethnicity in India*, 199.

68. In the colonial lexicon the term "jungle" denoted "a forest; a thicket; a tangled wilderness." Yule and Burnell, *Hobson-Jobson*, accessed July 24, 2019, https://dsalsrv04.uchicago.edu/cgi-bin/app/hobsonjobson_query.py?qs=JUNGLE&searchhws=yes.

69. Skaria, *Hybrid Histories*, ix.

70. Scott, *The Art of Not Being Governed*. Bhukya uses Scott's thesis to argue that the deep history of the Gond polity and society needs to be situated in their resistance to and rebellion against centralized state formation in central and southern India (*The Roots of the Periphery*).

71. Grant, *The Gazetteer of the Central Provinces*, xli.

72. F. Dangerfield, "Appendix II," in Malcolm, *A Memoir of Central India*, 2:313–47; Malcolmson, "On the Fossils"; Carter, "Summary," 255–64; Medlicott, "On the Geological Structure."

73. Voysey, "On some Petrified Shells," 194.

74. Malcolmson, "On the Fossils."

75. Sykes, "On a Fossil Fish," 272.

76. Malcolmson, "On the Fossils," 96–97.

77. Malcolmson, "On the Fossils," 97.

78. Grant, *The Gazetteer of the Central Provinces*, xl.

79. Sleeman, *Rambles and Recollections*, 1:127.

80. Sleeman, *Rambles and Recollections*, 1:127–28.

81. Sleeman, *Rambles and Recollections*, 1:125–27. Quotation on 125.

82. Grant, *The Gazetteer of the Central Provinces*, xl.

83. George Smith, *Stephen Hislop*, 68–69.

84. Hislop and Hunter, "On the Geology and Fossils," 348–49.

85. Hislop and Hunter, "On the Geology and Fossils," 348.

86. Hislop to Hunter, January 13, 1851, in Hislop manuscripts, 8958, 121; and Diary entries for January 3, 1855, and January 13, 1857, in Hislop manuscripts, 8960, 72, 103. Also see, George Smith, *Stephen Hislop*, 253, 263. Virapa belonged to the Lingayat community, a Vaishnav sect of southern India, belonging to the lower caste, often with distinct religious and cultural identities.

87. George Smith, *Stephen Hislop*, 264–65.

88. Hislop, "Geology of the Nagpur State," 58–76.

89. Hislop, "Geology of the Nagpur State," 58–76.

90. Thomas Oldham to Hislop, Calcutta, June 16, 1855, Hislop Manuscripts, 8957, 112.

91. Edinburgh, December 7, 1859, in George Smith, *Stephen Hislop*, 286.

92. Hislop, "Geology of the Nagpur State"; Hislop and Hunter, "On the Geology."

93. Hislop and Hunter, "On the Geology," 361.

94. George Smith, *Stephen Hislop*, 270.

95. Hislop and Hunter, "On the Geology," 382–83.

96. Slater, "Fossil Focus."

97. Although the *Glossopteris* has been later described as a woody shrub or tree, in the nineteenth century it was considered a fern because of the nature of the impressions of its leaves on the rocks.

98. Hislop and Hunter, "On the Geology," 382–83.

99. George Smith, *Stephen Hislop*, 269–70.

100. W. T. Blanford. "On the Geology of the Taptee," 235–37.

101. Grant, *The Gazetteer of the Central Provinces*, xlvi.

102. Newbold, "Summary of the Geology of Southern India," 252.

103. W. T. Blanford, "On the Geology of the Taptee," 235–37.

104. W. T. Blanford, "On the Geology of the Taptee," 235–37.

105. George Smith, *Stephen Hislop*, 269–70.

106. George Smith, *Stephen Hislop*, 269–70.

107. W. T. Blanford, "On the Geology of the Taptee," 235–37.

108. Foote, *Collection*, 36.

109. Foote, *Collection*, 60.

110. Foote, *Collection*, 184.

111. George Smith, *Stephen Hislop*, 340–43.
112. George Smith, *Stephen Hislop*, 340–43.
113. George Smith, *Stephen Hislop*, 271.
114. Newbold, "Summary of the Geology of Southern India," 257.
115. W. T. Blanford, *Observations*, 196. Near the British military camp at Buya, near Hintalo, he found the plains covered with black soil very similar to the *regur* of the Deccan (69).
116. "Specimens from Abyssinia from W. T. Blanford, 1869," DF PAL/105/2/9, Blanford correspondences.
117. W. T. Blanford, "On the Traps."
118. W. T. Blanford, *Observations*, 184–85.
119. W. T. Blanford, "On the Fauna," 107. He later published a more detailed article: "The African element."
120. W. T. Blanford, "Note on the geological age." For the role of the Blanford brothers in the making of the Gondwanaland theory, see Leviton and Aldrich, "The Impact of Travels on Scientific Knowledge."
121. H. F. Blanford, "On the age and correlation."
122. H. F. Blanford, "On the age and correlation," 534.
123. Suess, *Face of the Earth*, 1:595–97.
124. W. T. Blanford to Arthur Smith Woodward, September 25, 1890, Campden Hill, London, DF PAL/100/25/1, Blanford correspondences.
125. Stow, "On some Points in South-African Geology," particularly 546–48.
126. Stow, "On some Points in South-African Geology," 544.
127. Stow, "On some Points in South-African Geology," 546.
128. Stow, "On some Points in South-African Geology," 548.
129. Suess, *Face of the Earth*, 1:417–18.
130. Temple, "Note by the Editor on the Gond Songs," i–ii.
131. Temple, *Papers related to the Aboriginal Tribes*, 3–4.
132. "The Birth, Life, and Death of Lingo," in Temple, *Papers related to the Aboriginal Tribes*, part 3, 12.
133. Chatterton, *Story of Gondwana*, 161.
134. Russell, *The Tribes and Castes of the Central Provinces*, 3:48–52.
135. Gregory McNamee, *The Girl Who Made Stars*. See in particular, the introduction, 9–16. Quotation on 14.
136. Spillman, *British Colonial Realism in Africa*, 202–3.
137. Rachel King, "'A Loyal Liking for Fair Play,'" 410.
138. Skaria, "Some Aporias of History."
139. See for example, Wessels "The Creation of the Eland"; Byrne, "Deep Nation."
140. Castro, "Perspectivism and Multinaturalism."
141. For a more detailed study of Castro's perspectivism, see Vanzolini and Cesarino, "Perspectivism."
142. Chatterton, *Story of Gondwana*, 222–23.
143. Forsyth, *The Highlands of Central India*, 76.
144. Nite, "Worshipping the Colliery-Goddess."
145. Mehta, *Gonds of the Central Indian Highlands*, 1:126–27.
146. Elwin, *The Muria and Their Ghotul*, 236.
147. Elwin, *The Muria and Their Ghotul*, 237.
148. Patankar, "'Gondwana'/'Gondwanaland.'"
149. For a detailed study of the Gondwana movement and Punem, see Akash K. Prasad, "Gondwana Movement in Post-colonial India."

150. Laurasia was also known as Angaraland, referring to the region from the Ural Mountains to the Pacific coast.

151. Quoted in Patankar, "'Gondwana' / 'Gondwanaland,'" 39.

152. Connell, *Southern Theory.*

153. Connell, "The Shores of the Southern Ocean," 61.

154. Witzel, *The Origins of the World's Mythologies*, 216–18.

155. Stafford, "Geological Surveys," 10.

156. The most prominent writer in this field on colonial India is Deepak Kumar. See his *Science and the Raj*; "The Evolution of Colonial Science in India"; "Patterns of Colonial Science"; "Science, Resources, and the Raj." Also see Sangwan, *Science, Technology and Colonisation.* For a general exposition of this thesis of science and colonization, see Stafford, *Scientist of Empire.*

157. Sangwan, "Reordering the Earth"; Sangwan, "From Natural History to History of Nature." Sangwan draws inspiration from Richard Grove's thesis that suggested that colonial science engendered environmentalism and ecological consciousness more than the colonial exploitation of resources. Grove, *Green Imperialism.*

158. Moore, "The Capitalocene," 621. Also see Moore, *Anthropocene or Capitalocene?*

159. Moore, "The Capitalocene," 603–20. Also see Moore, *Capitalism in the Web of Life.*

160. Hobday, "Gondwana Coal Basins of Australia and South Africa," 219–20.

161. Fermor, "Gondwanaland," 491–94.

162. A recent article has stressed the role of "geological agency" in the hydrocarbon emissions that define the Anthropocene. See Walker and Johnson, "On Mineral Sovereignty."

163. Atherstone, "Nuggets of the Gouph," 1–3.

164. Atherstone, "Nuggets of the Gouph," 4. Gouph was that part of Karoo where the gold mines were located.

165. For South African and Australian histories of deep time that suggest similar possibilities, see Dubow, "Earth History, Natural History, and Prehistory"; and Tom Griffiths, "Deep Time and Australian History."

Conclusion

1. For a review of some of the most important recent writings on these themes, see Giraud, "The Planetary Is Political."

2. McPhee, *Basin and Range.*

3. Tom Griffiths, "Travelling in Deep Time."

4. Martin J. Rudwick, "Making History a Science," in his *Earth's Deep History*, 1–30.

5. Although innovative in its approach and stimulating and insightful in its attempt at writing deep history collectively with historians, anthropologists, and archaeologists, Shryock and Smail's *Deep History* falls within the older tradition. The book accepts prehistory as a natural deepening of "shallow" history that started in the nineteenth century. As has been suggested earlier, in this collaborative project, historians have provided historical clues to "paleohistorians" but have not questioned their methods and tools, which historians otherwise have done with their own methods. Shryock and Smail present singular "Hominin" species histories of the "body," which is neither racialized nor gendered (chapter 3); of "food," access to which is not politically and economically determined (chapter 6); and of "migration" that does not feature histories of political conflict and marginalization (chapter 8).

6. Tom Griffiths, "Travelling in Deep Time"; Bulstrode, "The Industrial Archaeology of Deep Time."

7. See for example, Wessels, "The Creation of the Eland"; Byrne, "Deep Nation"; Somerville, *Water in a Dry Land.*

8. Although, as Linda Andersson Burnett has shown, even in Europe, in the case of the Sami people, such marginalization is key to approaching the political content of their deep history ("Sami Indigeneity").

9. Here I am referring mostly to McGrath and Jebb, *Long History, Deep Time*.

10. McGrath and Jebb, *Long History, Deep Time*, 1–32. Quotation on 6.

11. Along with the works mentioned earlier by Ramaswamy, Banerjee, Tom Griffiths, and McGrath, there are Billy Griffiths, *Deep Time Dreaming*; van Sittert, "The Supernatural State"; Geertsema, "Imagining the Karoo Landscape"; Negi, " 'You Cannot Make a Camel Drink Water.' "

12. Masih, "India Orders 'Staggering' Eviction." The government of India has since instructed the court to stay the order.

13. Sethi, "SC Eviction Order."

14. Bandi, "Forest Rights Act"; "Question Raised in Lok Sabha."

15. Saravanan, "Political Economy".

16. "Indigenous Indian People Face Forest Eviction."

17. The term "green-black" is used in the Australian context to refer to the relationships between environmentalists and Aboriginals on various ecological and traditional rights issues.

18. Vincent and Neale, *Unstable Relations*.

19. John Vidal, "The Tribes Paying the Brutal Price of Conservation."

20. See, for example, Howlett and Lawrence, "Accumulating Minerals and Dispossessing Indigenous Australians."

21. Danino, "Aryans and the Indus Civilization." Recently, archaeological excavations carried out by the Indian National Institute of Ocean Technology in the Gulf of Cambay supposedly found ancient cities and remains at a depth of 20–40 meters in the seabed, which have been described as early "Aryan" settlements. See Witzel, "Rama's Realm."

22. Saravanan, *Environmental History and Tribals*, 159–84.

23. For example, Mamdani sees the Rwandan genocide as a product of colonialism, which reignited the existing divide between the settler and the people of the soil (*When Victims Become Killers*, particularly chapters 2 and 3).

24. Cam Sharp Jones, "Colonial Ethnography."

25. Baruah, *Postfrontier Blues*, 26.

26. Kar, "When Was the Postcolonial?," 51–52.

Primary Sources

Archival Sources

United Kingdom

British Library (BL), London: Asia, Pacific & Africa Collections (APAC)
India Office Records and Private Papers
Board's Collections: IOR/F/4: 1620–1859
Bombay Proceedings: IOR/P/347/46
Letter books of Sir Richard Jenkins (1785–1853) Mss Eur E111–12; F34: 1807–1817
Natural History Museum, London
W. T. Blanford correspondences, Natural History Museum Archives (Blanford Manuscripts)
Edinburgh University Library Special Collections Division, Scotland
Papers of Charles Lyell, GB 0237
National Library of Scotland, Archives and Manuscripts Collections, Edinburgh, Scotland
Papers of Stephen Hislop, missionary in Nagpur (Hislop Manuscripts)

India

National Archive of India, New Delhi (NAI)
Survey of India papers

United States

Yale University, New Haven, CT
G. Evelyn Hutchinson Papers (MS 649). Manuscripts and Archives, Yale University Library

Contemporary Journals

The Asiatic Journal and Monthly Register for British and Foreign India, China and Australasia
Asiatick Researches
Calcutta Journal of Natural History
Calcutta Review
Geological Magazine
Gleanings in Science
Journal of the Asiatic Society of Bengal
Journal of the Bombay Branch of the Royal Asiatic Society
Journal of the Royal Asiatic Society of Great Britain and Ireland
Madras Journal of Literature and Science
Memoirs of the Geological Survey of India: Palaeontologia Indica, being figures and descriptions of the organic remains procured during the progress of the Geological Survey of India
Proceedings of the Asiatic Society of Bengal
Proceedings of the Zoological Society of London
Quarterly Journal of the Geological, Mining and Meteorological Society of India
Quarterly Journal of the Geological Society of London
The Quarterly Oriental Magazine, Review and Register
Records of the Geological Survey of India
Transactions of the Geological Society of London
Transactions of the Literary Society of Bombay
Transactions of the Literary Society of Madras
Transactions of the Royal Asiatic Society of Great Britain and Ireland

Published Works

The Ain i Akbari by Abul Fazl Allami, Translated from the Original Persian by Colonel H. S. Jarrett, Secretary and Member, Board of Examiners. 2 vols. Calcutta, 1891.

Almond, Philip C. "Druids, Patriarchs, and the Primordial Religion." *Journal of Contemporary Religion* 15 (2000): 379–94.

Amrith, Sunil S. *Crossing the Bay of Bengal: The Furies of Nature and the Fortunes of Migrants.* Cambridge, MA: Harvard University Press, 2013.

App, Urs. *The Birth of Orientalism.* Philadelphia: University of Pennsylvania Press, 2010.

Armitage, Simon J., Simon J. Armitage, Sabah A. Jasim, Anthony E. Marks, Adrian G. Parker, Vitaly I. Usik, and Hans-Peter Uerpmann. "The Southern Route 'Out of Africa': Evidence for an Early Expansion of Modern Humans into Arabia." *Science* 331 (2011): 453–56.

Arnold, David. *The Tropics and the Traveling Gaze: India, Landscape, and Science 1800–1856.* Delhi: Permanent Black, 2005.

Asante, Molefi Kete. *Kemet, Afrocentricity, and Knowledge.* Trenton, NJ: Africa World Press, 1990.

Asif, Manan Ahmed. *A Book of Conquest: The Chachnama and Muslim Origins in South Asia.* Cambridge, MA: Harvard University Press, 2016.

Asif, Manan Ahmed. "The Long Thirteenth Century of the Chachnama." *Indian Economic and Social History Review* 49 (2012): 459–91.

Atherstone, W. G. "Nuggets of the Gouph." *Cape Monthly Magazine*, July 1873, 1–5.

Atkinson, Edwin T. *Statistical, Descriptive and Historical Account of the North-western Provinces of India.* Allahabad: North-Western Provinces and Oudh Government, 1874.

Atsushi, Akera. "Constructing a Representation for an Ecology of Knowledge: Methodological Advances in the Integration of Knowledge and Its Various Contexts." *Social Studies of Science* 37 (2007): 413–41.

Badam, G. L. "Colossochelys Atlas, a Giant Tortoise from the Upper Siwaliks of North India." *Bulletin of the Deccan College Post-Graduate and Research Institute* 40 (1981): 149–53.

Baird Smith, Richard. *Canals of Irrigation in the North Western Provinces of India.* Calcutta: Calcutta Review, 1849.

Baird Smith, Richard. "On the Structure of the Delta of the Ganges, as Exhibited by the Boring Operations in Fort William, AD 1836–40." Calcutta Journal of Natural History 1 (1840): 324–43.

Baird Smith, Richard. *A Short Account of the Ganges Canal: With a Description of the Some of the Principal Works.* Roorkee: Thomason College Press, 1870.

Baker, W. E., and H. M. Durand. "Sub-Himalayan Fossil Remains of the Dadupur Collection. Quadrumana." *Journal of the Asiatic Society of Bengal* 5 (1836): 739–41.

Bakker, Hans. "Ayodhyā: A Hindu Jerusalem; An Investigation of 'Holy War' as a Religious Idea in the Light of Communal Unrest in India." *Numen* 38 (1991): 80–109.

Baldaeus, Philippus. *A True and Exact Description of the Most Celebrated East-India Coasts of Malabar and Coromandel, as Also of the Isle of Ceylon . . . Also a Most Circumstantial and Compleat Account of the Idolatry of the Pagans in the East Indies . . . Translated from the High Dutch.* 1745.

Balfour, Edward. *Cyclopædia of India and of Eastern and Southern Asia, Commercial, Industrial and Scientific: Products of the Mineral, Vegetable and Animal Kingdoms, Useful Arts and Manufactures.* Madras: Scottish Press, 1857 (1873).

Ball, Valentine. *Jungle Life in India: Or, the Journeys and Journals of an Indian Geologist.* London: Thos. De la rue, 1880.

Ball, Valentine. "Stone Monuments in the District of Singhbhum-Chota Nagpur." *Indian Antiquary* 1 (1872) 291–92.

Bandi, Madhusudan. "Forest Rights Act: Towards the End of Struggle for Tribals?" *Social Scientist* 42 (2014): 63–81.

Banerjee, Prathama. *Politics of Time: "Primitives" and History-Writing in a Colonial Society*. Oxford: Oxford University Press, 2006.

Barnett, L. D. "A Seal of Sri-vadra." *Journal of the Royal Asiatic Society of Great Britain and Ireland* (April 1914): 401–2.

Barrell, Joseph. "Probable Relations of Climatic Change to the Origin of the Tertiary Ape-Man." *Scientific Monthly* 4, no. 1 (1917): 16–26.

Baruah, Sanjib. *Postfrontier Blues: Toward a New Policy Framework for Northeast India*. Policy Studies 33. Washington, DC: East-West Center, 2007.

Bashford, Alison. "The Anthropocene Is Modern History: Reflections on Climate and Australian Deep Time." *Australian Historical Studies* 44 (2013): 341–49.

Bashford, Alison. "Deep Genetics: Universal History and the Species." *History and Theory* 57 (2018): 313–22.

Bassuk, Daniel. *Incarnation in Hinduism and Christianity: The Myth of the God-Man*. Basingstoke: Macmillan Press, 1987.

Bates, Crispin. "Race, Caste and Tribe in Central India: The Early Origins of Anthropometry." In *The Concept of Race in South Asia*, edited by Peter Robb, 219–59. Delhi: Oxford University Press, 1997.

Batuta, Ibn. *The Travels of Ibn Batuta: Translated with Notes, by Samuel Lee*. London: Oriental Translation Committee, 1829.

Baviskar, Amita. *In the Belly of the River: Tribal Conflicts over Development in the Narmada Valley*. Delhi: Oxford University Press, 1995.

Bayly, C. A. *Empire and Information: Intelligence Gathering and Social Communication in India, 1780*. Cambridge: Cambridge University Press, 2000.

Bayly, C. A. *Rulers, Townsmen and Bazaars: North Indian Society in the Age of British Expansion, 1770–1870*. Delhi: Oxford University Press, 1992.

Beinart, William. *The Rise of Conservation in South Africa: Settlers, Livestock, and the Environment 1770–1950*. Oxford: Oxford University Press, 2003.

Bergen, Sadie. "What Are You? Historians Confront Race, Genealogy, and Genetics." Accessed July 24, 2019, https://www.historians.org/publications-and-directories/perspectives-on-history/february-2018/what-are-you-historians-confront-race-genealogy-and-genetics.

Béteille, André. "Caste and Citizen." *Science and Culture* 77 (2011): 83–90.

Bhadra, K., A. K. Gupta, and J. R. Sharma, "Saraswati Nadi in Haryana and Its Linkage with the Vedic Saraswati River—Integrated Study Based on Satellite Images and Ground Based Information." *Journal of the Geological Society of India* 73 (2009): 273–88.

Bhaskar, Roy. *The Possibility of Naturalism: A Philosophical Critique of the Contemporary Human Sciences*. Atlantic Highlands, NJ: Humanities Press, 1979.

Bhattacharya, Neeladri. "Myth, History and the Politics of Ramjanmabhumi." In *Anatomy of a Confrontation: Ayodhya and the Rise of Communal Politics in India*, edited by Sarvepalli Gopal, 122–40. London: Zed Books, 1993.

Bhukya, Bhangya. "Enclosing Land, Enclosing *Adivasis*: Colonial Agriculture and *Adivasis* in Central India, 1853–1948." *Indian Historical Review* 40 (2013): 93–116.

Bhukya, Bhangya. *The Roots of the Periphery: A History of the Gonds of Deccan India*. New Delhi: Oxford University Press, 2017.

Bigg, Charlotte, David Aubin, and Philipp Felsch. "Introduction: The Laboratory of Nature—Science in the Mountains." *Science in Context* 22 (2009): 311–21.

Bird, R. T. *Bones for Barnum Brown: Adventures of a Dinosaur Hunter*. Fort Worth, TX: Christian University Press, 1985.

Bishop, Peter. *The Myth of Shangri-La: Tibet, Travel Writing and the Western Creation of Sacred Landscape*. London: Athlone, 1989.

Blair, Ann. "Mosaic Physics and the Search for a Pious Natural Philosophy in the Late Renaissance." *Isis* 91 (2000): 32–58.

Blair, Ann. *The Theater of Nature—Jean Bodin and Renaissance Science*. Princeton, NJ: Princeton University Press, 2017.

Blanford, Henry F. "On the age and correlation of the Plant-bearing series of India, and the former existence of an Indo-Oceanic continent." *Quarterly Journal of the Geological Society of London* 31 (1875): 519–54.

Blanford, Henry F. *The Rudiments of Physical Geography for the Use of Indian Schools, Etc.* 2nd ed. Calcutta: Thacker, Spink, 1874.

Blanford, Henry F., and W. T. Blanford. "On the Geological Structure and Relations of the Talcheer Coal Field, in the District of Cuttack." *Memoirs of the Geological Survey of India* 1 (1859): 33–89.

Blanford, W. T. "The African element in the fauna of India: A criticism of Mr. Wallace's views as expressed in the 'Geographical Distribution of Animals.'" *Annals and Magazine of Natural History* 18 (1876): 277–94.

Blanford, W. T. "Note on the geological age of certain groups comprised in the Gondwána series of India, and on the evidence they afford of distinct zoological and botanical terrestrial regions in ancient epochs." *Records of the Geological Survey of India* 9 (1876): 79–85.

Blanford, W. T. *Observations on the Geology and Zoology of Abyssinia, Made During the Progress of The British Expedition to that Country In 1867–68*. London: Macmillan, 1870.

Blanford, W. T. "On the Fauna of British India, and its relations to the Ethiopian and so-called Indian fauna." In "Notes and Abstracts of Miscellaneous Communications to the Sections." *Report thirty-ninth meeting of the British Association for the Advancement of Science held at Exeter in 1869*. London: John Murray, 1870, 107–8.

Blanford, W. T. "On the Geology of the Taptee and Lower Nerbudda valleys, and some adjoining Districts." *Memoirs of the Geological Survey of India* 6 (1869): 163–384.

Blanford, W. T. "On the Traps and Intertrappean Beds of Western and Central India." *Memoirs of the Geological Survey of India* 6 (1869): 137–62.

Blunt, James T. "A Description of the Cuttub Minar." *Asiatick Researches* 4 (1799): 323–28.

Blunt, James T. "Narrative of a Route from Chunarghur, to Yertnagoodum, in the Ellore Circar." *Asiatick Researches* 7 (1803): 57–169.

Bose, P. N. "Chhattisgar: Notes on its Tribes, Sects and Castes." *Journal of the Asiatic Society of Bengal* 59 (1891): 269–300.

Boule, Marcellin. *Fossil Men: Elements of Human Palaeontology*. Edinburgh: Oliver & Boyd, 1923.

Bowler, Peter. "From 'Savage' to 'Primitive': Victorian Evolutionism and the Interpretation of Marginalized Peoples." *Antiquity* 66 (1992): 721–29.

Braudel, Fernand. *The Mediterranean and the Mediterranean World in the Age of Philip II*. London: Collins, 1973.

Braudel, Fernand, Roselyne De Ayala, and Paule Braudel. *Les écrits de Fernand Braudel*. Vol. 2, *Les ambitions de l'histoire*. Paris: Fallois, 1997.

Briggs, John. *History of the Rise of the Mahomedan Power in India, till the year A.D. 1612: Translated from the Persian of Mahomed Kasim Ferishta*. London: Longman, Rees, Orme, Brown, and Green, 1829.

Briggs, John. "Two Lectures on the Aboriginal Race of India, as distinguished from the Sanskritic or Hindu Race." *Journal of the Royal Asiatic Society of Great Britain & Ireland* 13 (1852): 275–309.

Brown, Barnum. "The Largest Known Land Tortoise: The Siwalik Hills of Northern India Yield the Complete Shell of a Fossil Tortoise That Weighed a Ton When Alive." *Natural History: The Journal of the American Museum of Natural History* 31 (1931): 183–87.

Brown, C. M. "Colonial and Post-colonial Elaborations of Avataric Evolutionism." *Zygon* 42 (2007): 715–48.

Brown, C. M. *Hindu Perspectives on Evolution: Darwin, Dharma and Design*. Abingdon: Routledge, 2012.

Brown, Joyce. "A Memoir of Colonel Sir Proby Cautley, FRS, 1802–1871, Engineer and Palaeontologist." *Notes and Records of the Royal Society of London*, 34 (1980): 185–225.

Brown, Joyce. "Smith, Robert, Colonel (1787–1873)." In *A Biographical Dictionary of Civil Engineers in Great Britain and Ireland (1500–1830)*, vol. 1, edited by A. W. Skempton, 637. London: Thomas Telford, 2002.

Browne, Janet. "Missionaries and the Human Mind: Charles Darwin and Robert Fitzroy." In *Darwin's Laboratory: Evolutionary Theory and Natural History in the Pacific*, edited by R. MacLeod and P. F. Rehbock, 263–82. Honolulu: University of Hawaii Press, 1994.

Bryant, Edwin. *The Quest for the Origins of Vedic Culture: The Indo-Aryan Migration Debate*. Oxford: Oxford University Press, 2001.

Bryant, Jacob. *A New System, or An Analysis of Ancient Mythology: Wherein an Attempt is Made to Divest Tradition of Fable; and to Reduce the Truth to Its Original Purity*. London: T. Payne; Elmsly; B. White; and J. Walter, 1780.

Buchan, Bruce, and Linda Andersson Burnett, eds. "Knowing Savagery: Humanity in the Circuits of Colonial Knowledge." *History of the Human Sciences* 32, no. 4 (2019): 3–134.

Bulstrode, Jenny. "The Industrial Archaeology of Deep Time." *British Journal for the History of Science* 49 (2016): 1–25.

Burkill, I. H. "Chapters on the History of Botany in India. I. From the Beginning to the Middle of Wallich's Service." *Journal of the Bombay Natural History Society* 51 (1954): 846–78.

Burnet, Thomas. *The Sacred Theory of the Earth: Containing an Account of the Original of the Earth, and of All the General Changes Which It hath already undergone, or Is to Undergo, till the Consummation of All Things*. London: R. Norton, 1691.

Burnett, Linda Andersson. "Sami Indigeneity in Nineteenth-Century Swedish and British Intellectual Debates." In *Companion to Indigenous Global History*, edited by Ann McGrath and Lynette Russell. London: Routledge, 2020 (forthcoming).

Burrow, Reuben. "A Proof that the Hindus had the Binomial Theorem." *Asiatick Researches* 2 (1790): 487–97.

Byrne, Denis. "Deep Nation: Australia's Acquisition of an Indigenous Past." *Aboriginal History* 20 (1996): 82–107.

Caldwell, Robert. *A Comparative Grammar of the Dravidian or South-Indian Family of Languages*. London, 1856.

Cannon, Garland Hampton, ed. *The Collected Works of Sir William Jones*. 13 vols. Richmond: Curzon Press, 1993.

Cannon, Garland Hampton, ed. *The Letters of Sir William Jones*. 2 vols. Oxford: Clarendon Press, 1970.

Cannon, Garland Hampton. *The Life and Mind of Oriental Jones: Sir William Jones, the Father of Modern Linguistics*. Cambridge: Cambridge University Press, 1990.

Carter, H. J. "Summary of the geology of India, between the Ganges, the Indus, and Cape Comorin." *Journal of the Bombay Branch of the Royal Asiatic Society* 5 (1854): 179–335.

Castro, Eduardo Viveiros De. "Perspectivism and Multinaturalism in Indigenous America." In *The Land Within: Indigenous Territory and the Perception of the Environment*, edited by Alexandre Surrallés and Pedro García Hierro, 36–75. Copenhagen: IWGIA, 2005.

Cautley, Proby T. "Discovery of an Ancient town near Behut, in the Doab." *Journal of the Asiatic Society of Bengal* 3 (1834): 43–44.

Cautley, Proby T. "Further Account of the Remains of an Ancient Town Discovered at Behat, near Seharanpur." *Journal of the Asiatic Society of Bengal* 3 (1834): 221–27.

Cautley, Proby T. *Ganges Canal: A disquisition on the heads of the Ganges and Jumna Canals, North-Western Provinces, in reply to strictures by Major-General Sir Arthur Cotton.* London: Printed for private circulation, 1864.

Cautley, Proby T. *Notes and Memoranda on the Eastern Jumna, or Doab Canal, North Western Provinces.* Calcutta: Bengal Military Orphan Press, 1845.

Cautley, Proby T. "On the Fossil Remains of Camilidae of the Sewalikhs." *Journal of the Asiatic Society of Bengal* 9 (1840): 620–23.

Cautley, Proby T. "On the Structure of the Sewalik hills, and the Organic remains found in them." *Transactions of the Geological Society of London* 5, series 2 (1840): 267–78.

Cautley, Proby T. "On the Use of Wells, &c. in Foundations; as practised by the Natives of the Northern Doab." *Journal of the Asiatic Society of Bengal* 8, part 1 (1840): 327–40.

Cautley, Proby T. *Report on the Central Doab Canal.* Allahabad: Allahabad Mission Press, 1841.

Cautley, Proby T. *Report on the Ganges Canal Works: From Their Commencement until the Opening of the Canal in 1854.* London: Smith, Elder, 1860.

Cautley, Proby T., and Hugh Falconer. *Fauna Antiqua Sivalensis: Being the Fossil Zoology of the Sewalik Hills, in the North of India.* London: Smith, Elder, 1846.

Cautley, Proby T., and Hugh Falconer. "Notice on the Remains of a Fossil Monkey from the Tertiary Strata of the Sewalik Hills in the North of Hindoostan." *Transactions of the Geological Society of London* 5, series 2 (1840): 499–504.

Cautley, Proby T., and Hugh Falconer. "On Additional fossil species of the order of Quadrumana from the Sewalik Hills." *Journal of the Asiatic Society of Bengal* 6, no. 1 (1837): 354–61.

Cautley, Proby T., and Hugh Falconer. "Sivatherium Gigantium. A New Found Ruminant Genus: From the Valley of the Markanda in the Shivalik Branch of Sub-Himalayan Mountains." *Asiatick Researches* 19 (1836): 1–24.

Census of India, 1931. Vol. 1, part 3. Simla: Government of India Press, 1935.

Chadha, Ashish. "Conjuring a River, Imagining Civilisation." *Contributions to Indian Sociology*, 45 (2011): 55–83.

Chakrabarti, Dilip K. *India: An Archaeological History: Palaeolithic Beginnings to Early Historic Foundations.* New Delhi: Oxford University Press, 1999.

Chakrabarti, Dilip K. "Robert Bruce Foote and Indian Prehistory." *East and West* 29 (1979): 11–26.

Chakrabarti, Pratik. "Beasts of Burden: Animals and Laboratory Research in Colonial India." *History of Science* 48 (2010): 125–52.

Chakrabarti, Pratik. *Medicine and Empire, 1600–1860.* Basingstoke: Palgrave, 2014.

Chakrabarti, Pratik. "Purifying the River: Pollution and Purity of Water in Colonial Calcutta." *Studies in History* 31 (2015): 178–205.

Chakrabarti, Pratik. *Western Science in Modern India, Metropolitan Methods, Colonial Practices.* Delhi: Permanent Black, 2004.

Chakrabarti, Pratik, and Joydeep Sen. "'The World Rests on the Back of a Tortoise': Science and Mythology in Indian History." *Modern Asian Studies* 50 (2016): 808–40.

Chakrabarty, Dipesh. "The Climate of History: Four Theses." *Critical Inquiry* 35 (2009): 197–222.

Chakrabarty, Dipesh. "Politics Unlimited: The Global *Adivasi* and the Debate about the Political." In *Indigeneity in India*, edited by T. Karlsson and T. B. Subba, 235–46. London: Kegan Paul, 2006.

Chakrabarty, Dipesh. "Postcoloniality and the Artifice of History: Who Speaks for 'Indian' Pasts?" *Representations*, 37 (1992): 1–26.

Chakrabarty, Ramakanta. *Vaisnavism in Bengal: 1486–1900.* Calcutta: Sanskrit Pustak Bhandar, 1985.

Chakraborty, Pratik. "Asiatic Society and Its Vision of Science: Metropolitan Knowledge in a Colonial World." *Calcutta Historical Journal* 21–22 (1999–2000): 1–32.

Chandrasekharam, D. "Geo-mythology of India." In *Myth and Geology*, edited by L. Piccardi and W. B. Masse, 29–37. Special Publication 273. London: Geological Society, 2007.

Chatterji, Suniti Kumar. "Contributions from Different Language-Culture Groups." In *The Cultural Heritage of India*, 1:76–92. Belur Math, Calcutta: Sri Ramakrishna Centenary Committee, 1936.

Chatterji, Suniti Kumar. "India and Polynesia: Austric Bases of Indian Civilization and Thought." In *Bharata-Kaumudi—Studies in Indology in Honour of Dr. Radhakumud Mhookerji*, 195–208. Allahabad: Indian Press, 1945.

Chatterji, Suniti Kumar. "The Internationalism of India." In *Nehru Abhinandan Granth*, 426–32. Calcutta: Aryavarta Prakashan Griha, 1949.

Chatterton, Eyre. *Story of Gondwana*, London: Sir I. Pitman & Sons, 1916.

Chippindale, Christopher. "The Invention of Words for the Idea of 'Prehistory.'" *Proceedings of the Prehistoric Society* 54 (1988): 303–14.

Cohen, Claudine. "Charles Lyell and the Evidences of the Antiquity of Man." In *Lyell: The Past Is the Key to the Present*, edited by Derek J. Blundell and C. Andrew Scott, 83–93. London: Geological Society, 1998.

Cohen, Claudine. *The Fate of the Mammoth: Fossils, Myth, and History*. Chicago: University of Chicago Press, 2002.

Cole, Henry Hardy. *The Architecture of Ancient Delhi: Especially the Buildings Around the Kutb Minar*. London: Arundel Society for Promoting the Knowledge of Art, 1872.

Colebrooke, Henry Thomas. Introductory note to "A Journey to Lake Mánasaróvara in Ún Dés, a province in Little Tibet," by William Moorcroft. *Asiatick Researches* 12 (1816): 375–76.

Colebrooke, Henry Thomas. "On the Heights of the Himálaya Mountains." *Asiatick Researches* 12 (1816): 251–85.

Colebrooke, Henry Thomas. "On the Religious Ceremonies of the Hindus, and the Bra'mens Especially." *Asiatick Researches* 7 (1803): 232–88.

Colebrooke, Henry Thomas. "On the Valley of the Setlej River, in the Himalaya Mountains, from the Journal of Captain A. Gerard." *Transactions of the Royal Asiatic Society of Great Britain and Ireland* 1 (1827): 343–80.

Coleman, Charles. *The Mythologies of the Hindus: With Plates*. London: Parbury, Allen, 1832.

Colvin, J. "On the Restoration of the Ancient Canals in the Delhi Territory." *Journal of the Asiatic Society of Bengal* 2 (1833): 105–27.

Connell, Raewyn. "The Shores of the Southern Ocean: Steps toward a World Sociology of Modernity, with Australian Examples." In *Worlds of Difference*, edited by Saïd Amir Arjomand and Elisa Pereira Reis, 58–72. London: Sage, 2013.

Connell, Raewyn. *Southern Theory: The Global Dynamics of Knowledge in Social Science*. Cambridge: Polity Press, 2007.

Constable, Philip. "Scottish Missionaries, 'Protestant Hinduism' and the Scottish Sense of Empire in Nineteenth- and Early Twentieth-Century India." *Scottish Historical Review* 86 (2007): 278–313.

Constituent Assembly Debates (Proceedings). Vol. 1, December 19, 1946. http://164.100.47.194/loksabha/writereaddata/cadebatefiles/C19121946.html.

"Continuation of the Route of Lieut. A. Burnes and Dr. Gerard, from Peshawar to Bokhara." *Journal of the Asiatic Society of Bengal* 2 (1833): 1–22.

Cook, Harold John. *Matters of Exchange: Commerce, Medicine, and Science in the Dutch Golden Age*. New Haven, CT: Yale University Press, 2007.

Cowper Reed, F. R. "Sedgwick Museum Notes." *Geological Magazine* 5 (1908): 256–61.

Crawfurd, John. *Journal of an Embassy from the Governor-General of India to the Court of Ava in the Year 1827, with an Appendix Containing a Description of Fossil Remains by Professor Buckland and Mr. Clift.* London: Henry Colburn, 1829.

Cregan-Reid, Vybarr. *Discovering Gilgamesh: Geology, Narrative and the Historical Sublime in Victorian Culture.* Manchester: Manchester University Press, 2011.

Cregan-Reid, Vybarr. "The Gilgamesh Controversy: The Ancient Epic and Late-Victorian Geology." *Journal of Victorian Culture* 14 (2009): 224–37.

Cunningham, Alexander. *The Ancient Geography of India.* London: Trübner, 1871.

Cunningham, Alexander. "Coins of the Indo-Scythians." *Numismatic Chronicle and Journal of the Numismatic Society* 8 (1888): 199–248.

Cunningham, Alexander. *Inscriptions of Asoka.* Calcutta: Office of the Superintendent of Government Printing, 1877.

Dalton, Edward Tuite. *Descriptive Ethnology of Bengal.* Calcutta: Office of the Superintendent of Government Printing, 1872.

Dalton, Edward Tuite. "Rude Stone Monuments in Chutiá Nagpúr and Other Places." *Journal of the Asiatic Society of Bengal* 42, part 1 (1873): 112–19.

Damodaran, Vinita. "Colonial Constructions of the 'Tribe' in India: The Case of Chotanagpur." *Indian Historical Review* 33 (2006): 44–75.

Damodaran, Vinita. "The Politics of Marginality and the Construction of Indigeneity in Chotanagpur." *Postcolonial Studies* 9 (2006): 179–96.

Danino, Michel. "Aryans and the Indus Civilization: Archaeological, Skeletal, and Molecular Evidence." In *A Companion to South Asia in the Past*, edited by Subhash R. Walimbe and Gwen Robbins Schug, 205–24. West Sussex: John Wiley and Sons, 2016.

Danino, Michel. "Genetics and the Aryan Debate." *Puratattva, Bulletin of the Indian Archaeological Society* 36 (2005): 146–54.

Danino, Michel. *The Lost River: On the Trail of the Sarasvati.* New Delhi: Penguin Books, 2010.

Darwin, Charles. *Descent of Man.* London: John Murray, 1871.

Das, Asok Kumar. *Wonders of Nature: Ustad Mansur at the Mughal Court.* Mumbai: Marg, 2012.

Dasgupta, Sangeeta. "'Heathen Aboriginals,' 'Christian Tribes,' and 'Animistic Races': Missionary Narratives on the Oraons of Chhotanagpur in Colonial India." *Modern Asian Studies* 50 (2016): 437–78.

Davidson, Levette J. "White Versions of Indian Myths and Legends." *Western Folklore* 7 (1948): 115–28.

Dawson, Gowan, and Bernard Lightman, eds. *Victorian Scientific Naturalism: Community, Identity, Continuity.* Chicago: University of Chicago Press, 2014.

De Almeida, Hermione, and George H. Gilpin. *Indian Renaissance: British Romantic Art and the Prospect of India.* London: Ashgate, 2005.

Dean, Edmund. "On the Fossil Bones of the Jamna River." *Journal of the Asiatic Society of Bengal* 4 (1835): 495–500.

Dean, Edmund. "On the Strata of the Jumna Alluvium, as exemplified in the Rocks and Shoals lately removed from the bed of the River; and of the sites of the Fossil Bones discovered therein." *Journal of the Asiatic Society of Bengal* 4 (1835): 261–78.

Descriptive sketch of the Sirhind Canal, commenced 1869, opened 1882. Lahore: Public Works Department, 1882.

de Secondat, Charles. *The Spirit of Laws*, trans. Thomas Nugent. Kitchener: Batoche Books, 1752/2001.

de Terra, Hellmut. "Pictorial History of Early Man Is Found Hewn in Rock in India." *New York Times*, December 16, 1935, 1.

de Terra, Hellmut. "Preliminary Report on the Yale North India Expedition." *Science* 26 (1933): 497–500.

de Terra, Hellmut. "Stone Age Tools Found in India." *New York Times*, January 12, 1936, 1.

Devji, Faisal. "India in the Muslim Imagination: Cartography and Landscape in 19th Century Urdu Literature." *South Asia Multidisciplinary Academic Journal*, no. 10 (2014): 1–17. Accessed July 24, 2019, https://doaj.org/article/a5b309cfd3ff4935b58486beb7f8eee4.

Dixon, R. B. *The Racial History of Man*. New York: Scribner, 1923.

Dow, Alexander. *The History of Hindostan: from the earliest account of time, to the death of Akbar; translated from the Persian of Mahammud Casim Ferishta of Delhi*. 3 vols. London: T. Becket, 1768.

Driberg, J. G., and H. J. Harrison. *Narrative of a Second visit to the Gonds of the Nurbudda Territory. With a grammar and vocabulary of their language*. Calcutta: Calcutta Diocesan Committee, College Press, 1849.

Driver, Felix. "Yule, Sir Henry." *Oxford Dictionary of National Biography*, accessed July 24, 2019, https://doi.org/10.1093/ref:odnb/30291.

Dubow, Saul. *A Commonwealth of Knowledge: Science, Sensibility, and White South Africa, 1820–2000*. Oxford: Oxford University Press, 2006.

Dubow, Saul. "Earth History, Natural History, and Prehistory at the Cape, 1860–1875." *Comparative Studies in Society and History* 46 (2004): 107–33.

Dubow, Saul. *Scientific Racism in Modern South Africa*. Cambridge: Cambridge University Press, 1995.

du Toit, Alexander L. *Our Wandering Continents: An Hypothesis of Continental Drifting*. London: Oliver and Boyd, 1937.

Eck, Diana L. *India: A Sacred Geography*. New York: Harmony, Random House, 2012.

Edelmann, Jonathan B. *Hindu Theology and Biology: The Bhāgavata Purāṇa and Contemporary Theory*. Oxford: Oxford University Press, 2012.

Edney, Matthew H. *Mapping an Empire: The Geographical Construction of British India, 1765–1843*. Chicago: University of Chicago Press, 1997.

Elwin, Verrier. *The Aboriginals*. Bombay: Oxford University Press, 1943.

Elwin, Verrier. *The Muria and Their Ghotul*. Bombay: Oxford University Press, 1947.

Ernst, Carl. "India as a Sacred Islamic Land." In *Religions of India in Practice*, edited by Donald S. Lopez, 556–63. Princeton, NJ: Princeton University Press, 1995.

Etter, Anne-Julie. "Antiquarian Knowledge and Preservation of Indian Monuments at the Beginning of the Nineteenth Century." In Sengupta and Ali, *Knowledge Production, Pedagogy*, 75–95.

Ewer, Walter. "An Account of the Inscriptions on the Cootub Minar, and on the Ruins in its Vicinity." *Asiatick Researches* 14 (1822): 480–88.

Falconer, Hugh. "Abstract of a Discourse, by Da. Falconer, on the Fossil Fauna of the Sewalik Hills." *Journal of the Royal Asiatic Society of Great Britain and Ireland* 8 (1846): 105–11.

Falconer, Hugh. "Description of some Fossil Remains of Dinotherium, Giraffe, and other Mammalia, from the Gulf of Cambay, Western Coast of India, chiefly from the Collection presented by Captain Fulljames, of the Bombay Engineers, to the Museum of the Geological Society." In *Geological Papers on Western India, including Cutch, Sinde, and the South-East Coast of Arabia*, by Henry J. Carter, 475–90. Bombay: Government of Bombay, 1857.

Falconer, Hugh. *Descriptive Catalogue of the Fossil Remains of Vertebrata from the Sewalik Hills, the Nerbudda, Perim Island, &c. in the Museum of the Asiatic Society of Bengal*. Calcutta: C. B. Lewis, 1859.

Falconer, Hugh. "Official report of Expedition to Cashmeer and Little Tibet, in 1837–38." In *Palæontological Memoirs and Notes of the Late Hugh Falconer*, ed. Murchison, 1:557–76. London: Robert Hardwicke, 1868.

Falconer, Hugh. "On the Asserted Occurrence of Human Bones in the ancient Fluviatile Deposits of the Nile and Ganges; with comparative Remarks on the Alluvial Formation of the two Valleys." *Quarterly Journal of the Geological Society of London* 21 (1865): 372–89.

Falconer, Hugh. "Primeval Man, and his Contemporaries." In Murchison, *Palæontological Memoirs*, 2:570–600.

Falconer, Hugh. "Report of Progress in the Brixham Cave, 1858." In Murchison, *Palæontological Memoirs*, 2:491–97.

Farrell, Gerry, and Neil Sorrell. "Colonialism, Philology, and Musical Ethnography in Nineteenth-Century India: The Case of S. W. Fallon." *Music and Letters* 88 (2007): 107–12.

Feldhaus, Anne. *Connected Places: Region, Pilgrimage, and Geographical Imagination in India.* Basingstoke: Palgrave Macmillan, 2003.

Fermor, L. L. "Gondwanaland, a Former Southern Continent." *Proceedings of the Bristol Naturalists' Society* 4 (1944): 483–94.

Fleetwood, Lachlan. "'No Former Travellers Having Attained Such a Height on the Earth's Surface': Instruments, Inscriptions, and Bodies in the Himalaya, 1800–1830." *History of Science* 56 (2018): 3–34.

Flood, Gavin D. *An Introduction to Hinduism.* Cambridge: Cambridge University Press, 1996.

Flower, William Henry. *Essays on Museums and other Subjects Connected with Natural History.* London, Macmillan, 1898.

Fludd, Robert. *Philosophia Moysaica: In qua Sapientia Et Scientia Creationis Et Creaturarum Sacra Vereque Christiana Ad Amussim Et Enucleate Explicatur.* Goudæ, 1638.

Foote, Robert Bruce. *The Foote Collection of Indian Prehistoric and Protohistoric Antiquities.* Madras: Printed by the Superintendent, Government, 1914.

Foote, Robert Bruce. "On the Distribution of Stone Implements in Southern India." *Quarterly Journal of the Geological Society of London*, 24 (1868): 484–95.

Foote, Robert Bruce. *On the Occurrence of Stone Implements in lateritic formations in various parts of the Madras and North Arcot districts.* Madras: Geological Survey of India, 1865.

Forsyth, James. *The Highlands of Central India: Notes on Their Forests and Wild Tribes, Natural History, and Sports.* London: Chapman and Hall, 1871.

Foucault, Michel. *Order of Things: An Archaeology of the Human Sciences.* London: Routledge, 1989 (1966).

Fox, Cyril. "The Gondwana System and Related Formations." *Memoirs Geological Survey of India* 58 (1931): 1–241.

Fox, Lane Augustus. *Catalogue of the Anthropological Collection lent by Colonel Lane Fox for exhibition in the Bethnal Green Branch of the South Kensington Museum*, London: G. E. Eyre, 1874.

Frankel, Henry R. *The Continental Drift Controversy.* 4 vols. Cambridge: Cambridge University Press, 2012.

Fulljames, George. "Note on the discovery of Fossil Bones of Mammalia in Katiawar." *Journal of the Bombay Branch of the Royal Asiatic Society* 1, no. 1 (1841): 30–33.

Gamble, Clive. "Palaeolithic Society and the Release from Proximity: A Network Approach to Intimate Relations." *World Archaeology* 29 (1998): 426–49.

Gamble, Clive, and Moutsiou, Theodora. "The Time Revolution of 1859 and the Stratification of the Primeval Mind." *Notes & Records of the Royal Society* 65 (2011): 43–63.

Garzanti, E. "Stratigraphy and Sedimentary History of the Nepal Tethys Himalaya Passive Margin." *Journal of Asian Earth Sciences* 17 (1999): 805–27.

Gazetteer of the Delhi District, 1883–84. Reprint, New Delhi: Aryan Books International, 2010.

Geertsema, Johan. "Imagining the Karoo Landscape: Free Indirect Discourse, the Sublime and the Consecration of White Poverty." In *Literary Landscapes: From Modernism to Postcolonialism*, edited by Attie De Lange, Gail Fincham, Jeremy Hawthorn, and Jakob Lothe, 92–108. New York: Palgrave Macmillan, 2008.

Ghosh, Kaushik. "A Market for Aboriginality: Primitivism and Race Classification in the Indentured Labour Market of Colonial India." In *Subaltern Studies 10: Writings on South Asian History and Society*, edited by Gautam Bhadra, Gyan Prakash, and Susie Tharu, 8–48. Delhi: Oxford University Press, 1999.

Gillispie, Charles Coulston. *Genesis and Geology: A Study in the Relation of Scientific Thought, Natural Theology, and Social Opinion in Great Britain, 1790–1850*. New York: Harper Torch Book, 1959.

Gilmartin, David. "Scientific Empire and Imperial Science: Colonialism and Irrigation Technology in the Indus Basin." *Journal of Asian Studies* 53 (1994): 1127–49.

Giosan, Liviu, Peter D. Clift, Mark G. Macklin, Dorian Q. Fuller, Stefan Constantinescu, Julie A. Durcan, Thomas Stevens, Geoff A. T. Duller, Ali R. Tabrez, Kavita Gangal, et al. "Fluvial Landscapes of the Harappan Civilization." *Proceedings of the National Academy of Sciences of the United States of America* 109 (2012): E1688–94.

Giraud, Eva. "The Planetary Is Political," *BioSocieties* 14, no. 3 (2019): 472–81. https://doi.org/10.1057/s41292-019-00169-1

Gold, Meira. "Ancient Egypt and the Geological Antiquity of Man, 1847–1863." *History of Science* 57 (2019): 194–230.

Goodrum, Matthew R. "The History of Human Origins Research and Its Place in the History of Science: Research Problems and Historiography." *History of Science* 47 (2009): 337–57.

Gopal, Sarvepalli, Romila Thapar, Bipan Chandra, Sabyasachi Bhattacharya, Suvira Jaiswal, Harbans Mukhia, K. N. Panikkar, R. Champakalakshmi, Satish Saberwal, B. D. Chattopadhyaya, et al. "The Political Abuse of History: Babri Masjid-Rama Janmabhumi Dispute." *Social Scientist* 18 (1990): 76–81.

Gosling, David L. "Science and the Hindu Tradition: Compatibility or Conflict?" *Zygon* 47 (2012): 575–88.

Gould, Stephen J. "The Geometer of Race." *Discover Magazine*, November 1994. accessed July 23, 2019, http://discovermagazine.com/1994/nov/thegeometerofrac441.

Gould, Stephen J. *The Mismeasure of Man: Reviewed and Expanded*. New York: Norton, 1996.

Gould, Stephen J. *The Structure of Evolutionary Theory*. Cambridge, MA: Harvard University Press, 2002.

Grant, Charles, ed. *The Gazetteer of the Central Provinces of India*. 2nd ed. Bombay: Education Society's Press, 1870.

Greene, Mott T. *Alfred Wegener: Science, Exploration, and the Theory of Continental Drift*, Baltimore: Johns Hopkins University Press, 2015.

Greene, Mott T. *Geology in the Nineteenth Century: Changing Views of a Changing World*. Ithaca, NY: Cornell University Press, 1982.

Gregory, William King, G. E. Lewis, and Milo Hellman. *Fossil Anthropoids of the Yale-Cambridge India Expedition of 1935*. Washington, DC: Carnegie Institution, 1938.

Griffiths, Billy. *Deep Time Dreaming: Uncovering Ancient Australia*. Carlton, Vic: Schwartz, 2018.

Griffiths, Tom. "Deep Time and Australian History." *History Today* 51 (2001): 20–25.

Griffiths, Tom. "Travelling in Deep Time: La Longue Durée in Australian History." Accessed July 24, 2019. http://australianhumanitiesreview.org/2000/06/01/travelling-in-deep-timela-longue-dureein-australian-history/.

Grout, Andrew. "Geology and India, 1775–1805: An Episode in Colonial Science." *South Asia Research* 10 (1990): 1–18.

Grove, Richard H. *Green Imperialism: Colonial Expansion, Tropical Island Edens and the Origins of Environmentalism, 1600–1860*. Cambridge: Cambridge University Press, 1995.

Guha, Biraja Sankar. "An Outline of the Racial Ethnology of India." In *Indian Science Congress Association: Silver Jubilee Session, 1938; An Outline of the Field Sciences of India*, 125–40. Calcutta: Indian Science Congress Association, 1938.

Guha, Biraja Sankar. "Racial Affinities of the People of India." In *Census of India, 1931*, vol. 1, part 3, 2–22. Simla: Government of India Press, 1935.

Guha, Ramachandra. "The Prehistory of Community Forestry in India." *Environmental History* 6 (2001): 213–38.

Guha, Ramachandra. "Savaging the Civilised: Verrier Elwin and the Tribal Question in Late Colonial India." *Economic and Political Weekly*, 31 (1996): 2375–89.

Guha, Sudeshna. *Artefacts of History: Archaeology, Historiography and Indian Pasts*. New Delhi: Sage, 2015.

Guha, Sumit. *Environment and Ethnicity in India 1200–1991*. Cambridge: Cambridge University Press, 1999.

Guha, Sumit. "Lower Strata, Older Races, and Aboriginal Peoples: Racial Anthropology and Mythical History Past and Present." *Journal of Asian Studies* 57 (1998): 423–41.

Guha-Thakurta, Tapati. "Archaeology as Evidence: Looking Back from the Ayodhya Debate." Occasional Paper no. 159. Calcutta: CSSS, 1997.

Guha-Thakurta, Tapati. *Monuments, Objects, Histories—Institutions of Art in Colonial and Postcolonial India*. New York: Columbia University Press, 2004.

Gupta, Narayani. *Delhi between Two Empires 1803–1931: Society, Government and Urban Growth*. Delhi: Oxford University Press, 1981.

Gupta, Narayani. "From Architecture to Archaeology: The 'Monumentalising' of Delhi's History in the Nineteenth Century." In *Perspectives of Mutual Encounters in South Asian History, 1760–1860*, edited by Jamal Malik, 49–64. Leiden: Brill, 2000.

Habib, Irfan. "Imaging River Sarasvati: A Defence of Commonsense." *Social Scientist* 29 (2001): 46–74.

Hall, Martin, and C. H. Borland. "The Indian Connection: An Assessment of Hromnik's 'Indo-Africa.'" *South African Archaeological Bulletin* 37 (1982): 75–80.

Hallam, Tony. "The Great Revolution in the Earth Sciences in the Mid-twentieth Century." *Isis* 105 (2014): 410–12.

"Haryana Govt Pumps 100 Cusec Water to Revive 'Lost' Saraswati." *Indian Express*, August 6, 2016. Accessed July 23, 2019, http://indianexpress.com/article/india/india-news-india/haryana-govt-pumps-100-cusec-water-to-revive-lost-saraswati-2956608/.

Head, Lesley. *Second Nature: The History and Implications of Australia as Aboriginal Landscape*. Syracuse, NY: Syracuse University Press, 2000.

Herbert, J. D. "On the Organic Remains found in the Himalaya." *Gleanings in Science* 3 (1831): 265–72.

Higgins, Godfrey. *The Celtic Druids, or an Attempt to Shew that the Druids were the Priests of Oriental Colonies who Emigrated from India, and were the Introducers of the First or Cadmean System of Letters, and the Builders of Stonehenge, of Carnac, and of other Cyclopean works in Asia and Europe*. London: Rowland Hunter, 1829.

Hislop, Stephen. "Essay on the Hill Tribes of the Central Provinces." In "Note by the Editor on the Gond Songs," by Richard Temple. In *Papers related to the Aboriginal Tribes of Central Provinces Left in Mss by the Late Revd. Stephen Hislop, Missionary of the Free Church of Scotland at Nagpore*, edited by Richard Temple, part 3, 26–27. 1866.

Hislop, Stephen. "Geology of the Nagpur State." *Journal of the Bombay Branch of the Royal Asiatic Society* 5 (1857): 58–76.

Hislop, Stephen, and Robert Hunter. "On the Geology and Fossils of the Neighbourhood of Nágpur, Central India." *Quarterly Journal of the Geological Society of London* 11 (1855): 345–83.

Hobday, D. K. "Gondwana Coal Basins of Australia and South Africa: Tectonic Setting, Depositional Systems and Resources." *Geological Society Special Publication* 32, no. 1 (1987): 219–33.

Hodgson, J. A., and J. D. Herbert. "An Account of Trigonometrical and Astronomical Operations for Determining the Heights and Positions of the Principal Peaks of the Himalaya Mountains." *Asiatick Researches* 14 (1822): 187–372.

Holditch, Thomas. *Gates of India: Being an Historical Narrative*. London: Macmillan, 1910.

Holt, Frank Lee. *Thundering Zeus: The Making of Hellenistic Bactria*. Berkeley: University of California Press, 1999.

Hosagrahar, Jyoti. *Indigenous Modernities: Negotiating Architecture and Urbanism*. London: Routledge, 2005.

Howlett, C., and R. Lawrence. "Accumulating Minerals and Dispossessing Indigenous Australians: Native Title Recognition as Settler-Colonialism." *Antipode* 51 (2019): 818–37.

Hügel, Baron, and George Fulljames. "Recent Discovery of Fossil Bones in Perim Island, in the Cambay Gulph." *Journal of the Asiatic Society of Bengal* 5 (1836): 288–91.

Humboldt, Alexander de. *Personal Narrative of Travels to the Equinoctial Regions of the New Continent during the Years 1799–1804*, London: Longman Hurst Rees Orme and Brown, 1819.

Huxley, T. H. "On the Geographical Distribution of the Chief Modifications of Mankind." *Journal of the Ethnological Society of London* 2 (1870): 404–12.

Imperial Gazetteer of India. Vol. 26, *Atlas*. Oxford: Clarendon Press, 1909.

"Indigenous Indian People Face Forest Eviction." *Ecologist: Journal for the Post-Industrial Age* (April 10, 2019). Accessed July 24, 2019, https://theecologist.org/2019/apr/10/indigenous-indian-people-face-forest-eviction.

Introduction to "History Meets Biology." *American Historical Review* 119 (2014): 1492–99.

Jaffrelot, Christophe. "Hindu Nationalism and the (Not So Easy) Art of Being Outraged: The Ram Setu Controversy." In *Religion, Caste, and Politics in India*, edited by Christophe Jaffrelot, 305–23. New Delhi: Primus Books, 2010.

Jardine, Lisa. *Ingenious Pursuits: Building the Scientific Revolution*. London: Anchor Books, 1999.

Jazeel, Tariq. *Sacred Modernity: Nature, Environment, and the Postcolonial Geographies of Sri Lankan Nationhood*. Liverpool: Liverpool University Press, 2013.

Jenkins, Richard. *Report on the Territories of the Rajah of Nagpore: Submitted to the Supreme Government of India*. Calcutta: Government Gazette Press, 1827.

Jones, Cam Sharp. "Colonial Ethnography and Tribal Pasts in India, 1820–1900." PhD diss., University of Manchester, 2019.

Jones, William. "A Discourse on the Institution of a Society, for Inquiring into the History, Civil and Natural, the Antiquities, Arts, Sciences, and Literature of Asia." *Asiatick Researches* 1 (1788): ix–xvi.

Jones, William. "On the Chronology of the Hindus." *Asiatick Researches* 2 (1799): 111–47.

Jones, William. "On the Gods of Greece, Italy and India." *Asiatick Researches* 1 (1788): 221–75.

Jones, William. "On the Origin and Families of Nations." *Asiatick Researches* 3 (1792): 479–92.

Jones, William. "The Third Anniversary Discourse [on the Hindus], delivered 2 February 1786." *Asiatick Researches* 1 (1788): 415–31.

Josephson-Storm, Jason Ā. *Myth of Disenchantment: Magic, Modernity, and the Birth of the Human Sciences*. Chicago: University of Chicago Press, 2017.

"The Jumna Canals." In *Selections from the Records of Government, North Western Provinces*. 2:57–60. Secundra: Orphan Press, 1855.

Kalapura, Jose. "Constructing the Idea of Tibet: Missionary Explorations, 17–18th Centuries." *Proceedings of the Indian History Congress* 73 (2012): 1065–77.

Kapila, Shruti. "Race Matters: Orientalism and Religion, India and beyond c. 1770–1880." *Modern Asian Studies* 41 (2007): 471–513.

Kar, Bodhisattva. "When Was the Postcolonial? A History of Policing Impossible Lines." In *Beyond Counter-insurgency: Breaking the Impasse in Northeast India*, edited by Sanjib Baruah, 49–77. Oxford: Oxford University Press, 2009.

Karlsson, T., and T. B. Subba, eds. *Indigeneity in India*. London: Kegan Paul, 2006.
Keen, Ian. "The Anthropologist as Geologist: Howitt in Colonial Gippsland." *Australian Journal of Anthropology* 11 (2000): 78–97.
Kelly, Donald. "The Rise of Prehistory." *Journal of World History* 14 (2003): 17–36.
Kennedy, Kenneth A. R. *God-Apes and Fossil Men: Paleoanthropology of South Asia*. Ann Arbor: University of Michigan Press, 2000.
Kennedy, Kenneth A. R., and Russell L. Ciochon. "A Canine Tooth from the Siwaliks: First Recorded Discovery of a Fossil Ape?" *Human Evolution* 14 (1999): 231–53.
Kennet, Charles Egbert. "Observations on the Language of the Gonds, South of the Nerbudda." *Madras Journal of Literature and Science* 16 (1850): 33–54.
King, Rachel. "'A Loyal Liking for Fair Play': Joseph Millerd Orpen and Knowledge Production in the Cape Colony." *South African Historical Journal* 67 (2015): 410–32.
King, William. "Notice of Pre-Historic Burial Place with Cruciform Monoliths, near Mungapet in the Nizám's Dominions." *Journal of the Asiatic Society of Bengal* 46 (1877): 179–85.
Kious, W., and I. Robert Tilling. *This Dynamic Earth: The Story of Plate Tectonics*. Washington DC: US Geological Survey, 1994.
Knell, Simon J. *The Culture of English Geology, 1815–1851: A Science Revealed through Its Collecting*. Aldershot: Ashgate, 2000.
Knight, Melvin M. "The Geohistory of Fernand Braudel." *Journal of Economic History* 10 (1950): 212–16.
Koch, E. "Netherlandish Naturalism in Imperial Mughal Painting." *Apollo* 152 (2000): 29–37.
Kumar, Deepak. "The Evolution of Colonial Science in India: Natural History and the East India Company." In *Imperialism and the Natural World*, edited by John M. Mackenzie, 51–66. Manchester: Manchester University Press, 1990.
Kumar, Deepak. "Patterns of Colonial Science." *Indian Journal of History of Science* 15 (1980): 105–13.
Kumar, Deepak. *Science and the Raj, 1857–1905*. Delhi: Oxford University Press, 1995.
Kumar, Deepak. "Science, Resources, and the Raj: A Case Study of Geological Works in the Nineteenth Century India." *Indian Historical Review* 10 (1983–84): 6–89.
Kumar, Sunil. "Qutb and Modern Memory." In *The Partitions of Memory: The Afterlife of the Division of India*, edited by Suvir Kaul, 140–82. Delhi: Permanent Black, 2001.
Lahiri, Nayanjot. "Archaeology and Identity in Colonial India." *Antiquity* 74 (2000): 686–92.
Lahiri, Nayanjot, ed. *The Decline and Fall of the Indus Civilization*. Delhi, Bangalore: Permanent Black, 2000.
Lahiri, Nayanjot. "Living Antiquarianism in India." In *World Antiquarianism: Comparative Perspectives*, edited by Alain Schnapp, Lothar von Falkenhausen, Peter N. Miller, and Tim Murray, 423–38. Los Angeles: Getty Research Institute, 2014.
Lake, Edward. *Sir Donald McLeod*. London: Religious Tract Society, 1873.
Lal, B. B. "Aryan Invasion of India: Perpetuation of a Myth." In *The Indo-Aryan Controversy: Evidence and Inference in Indian History*, edited by Edwin Bryant and Laurie L. Patton, 50–74. London: Routledge, 2004.
"The Late Mr. H. W. Voysey." *Quarterly Oriental Magazine, Review and Register* 1 (1824): cxiii.
Latour, Bruno. *We Have Never Been Modern*. Translated by Catherine Porter. New York: Harvester Wheatsheaf, 1993.
Leask, Nigel. "Francis Wilford and the Colonial Construction of Hindu Geography, 1799–1822." In *Romantic Geographies: Discourses of Travel, 1775–1844*, edited by Amanda Gilroy, 204–22. Manchester: Manchester University Press, 2000.
Lenoir, Timothy. "The Disciplines of Nature and the Nature of Disciplines." In *Knowledges: Historical and Critical Studies in Disciplinarity*, edited by Ellen Messer-Davidow, David R. Shumway, and David Sylvan, 70–102. Charlottesville: University of Virginia Press, 1993.

Leopold, Joan. "British Applications of the Aryan Theory of Race to India, 1850–1870." *English Historical Review* 89 (1974): 578–603.

Leviton, Alan E., and Michele L. Aldrich. "Contributions of the Geological Survey of India, 1851–1890, to the Concept of Gondwána-Land." *Earth Sciences History* 31 (2012): 247–69.

Leviton, Alan E., and Michele L. Aldrich. "The Impact of Travels on Scientific Knowledge: William Thomas Blanford, Henry Francis Blanford, and the Geological Survey of India, 1851–1889." Supplement 2, *Proceedings of the California Academy of Sciences* 55, no. 9 (2004): 117–37.

Lewis, G. Edward. "Preliminary Notice of New Man-Like Apes from India." *American Journal of Science* 27, no. 159 (1934): 161–81.

Lightman, Bernard V. "Huxley and Scientific Agnosticism: The Strange History of a Failed Rhetorical Strategy." *British Journal for the History of Science* 35 (2002): 271–89.

Livingstone, David N. *Adam's Ancestors: Race, Religion and the Politics of Human Origins*. Baltimore: Johns Hopkins University Press, 2008.

Livingstone, David N. "The Preadamite Theory and the Marriage of Science and Religion." *Transactions of the American Philosophical Society* 82, no. 3 (1992): 1–78.

Livingstone, David N., and Charles W. J. Withers, ed. *Geography and Enlightenment*. Chicago: University of Chicago Press, 1999.

Lloyd, William, and Alexander Gerard. *Narrative of a Journey from Caunpoor to the Boorendo Pass: In the Himalaya Mountains Viâ Gwalior, Agra, Delhi, and Sirhind*. London: J. Madden, 1840.

Locke, John. *An Essay concerning Human Understanding*. London: Tho. Basset, 1690.

The London, Edinburgh, and Dublin Philosophical Magazine and Journal of Science. Vol. 25. London: Taylor, 1844.

Lubbock, John. *Prehistoric Times as Illustrated by Ancient Remains, and the Manners and Customs of Modern Savages*. London: Williams and Norgate, 1865.

Lydekker, Richard. *Catalogue of the Remains of Siwalik Vertebrata contained in the Geological Department of the Indian Museum, Calcutta*. 2 vols. Calcutta: Superintendent of Government Printing, 1885–86.

Lydekker, Richard. "Further notices of Siwalik Mammalia." *Records of the Geological Survey of India* 12, no. 1 (1879): 33–52.

Lydekker, Richard. "Notices of Siwalik Mammals." *Records of the Geological Survey of India* 11, no. 1 (1878): 64–104.

Lydekker, Richard. "On the Land Tortoises of the Siwaliks." *Records of the Geological Survey of India* 22 (1889): 209–11.

Lyell, Charles. *Elements of Geology*. London: J. Murray, 1838.

Lyell, Charles. *The Geological Evidences of the Antiquity of Man*. London: John Murray, 1863.

Lyell, Charles. *A Manual of Elementary Geology: or, The Ancient Changes of the Earth and Its Inhabitants as Illustrated by Geological Monuments*. 5th ed. London: J. Murray, 1855.

Lyell, Charles. "On the Occurrence of works of human art in post-pliocene deposits." In "Notes and Abstracts." In *Report of the 29th meeting of the British Society for the Advancement of Science, Notices and Abstracts*, 93–95. London: John Murray, 1860.

Lyell, Charles. *Principles of Geology; or the Modern Changes of the Earth and its Inhabitants*. 7th ed. London: John Murray, 1847 (1830).

Mabbett, I. W. "The Symbolism of Mount Meru." *History of Religions* 23 (1983): 64–83.

Macaulay, Vincent, Catherine Hill, Alessandro Achilli, Chiara Rengo, Douglas Clarke, William Meehan, James Blackburn, Ornella Semino, Rosaria Scozzari, Fulvio Cruciani, et al. "Single, Rapid Coastal Settlement of Asia Revealed by Analysis of Complete Mitochondrial Genomes." *Science* 308 (2005): 1034–36.

Mackeson, F. "Journal of Captain C. M. Wade's voyage from Lodiana to Mithankot by the River Satlaj, on his mission to Lahore and Bhawalpur." *Journal of the Asiatic Society of Bengal* 6, no. 1 (1837): 169–217.

Majeed, Javed. *Ungoverned Imaginings: James Mill's The History of British India and Orientalism.* Oxford: Clarendon Press, 1992.

Malcolm, John. *A Memoir of Central India including Malwa and Adjoining Provinces.* 2nd ed. London: Kingsbury Parbury & Allen, 1824.

Malcolmson, John G. "On the Fossils of the Eastern Portion of the Great Basaltic District of India." *Madras Journal of Literature and Science* 12 (1840): 58–105.

Mamdani, Mahmood. *When Victims Become Killers: Colonialism, Nativism, and the Genocide in Rwanda.* Princeton, NJ: Princeton University Press, 2001.

Manias, Chris. "The Problematic Construction of 'Palaeolithic Man': The Old Stone Age and the Difficulties of the Comparative Method, 1859–1914." *Studies in History and Philosophy of Biological and Biomedical Sciences* 51 (2015): 32–43.

Manias, Chris. *Race, Science, and the Nation: Reconstructing the Ancient past in Britain, France and Germany.* London: Routledge, 2013.

Marshall, P. J., ed. *The British Discovery of Hinduism in the Eighteenth Century.* Cambridge: Cambridge University Press, 1970/2008.

Martineau, J. *The Life and Correspondence of the Right Hon Sir Bartle Frere.* London: John Murray, 1895.

Masih, Niha. "India Orders 'Staggering' Eviction of 1 Million Indigenous People: Some Environmentalists Are Cheering." *Washington Post*, February 22, 2019. Accessed July 23, 2019, https://www.washingtonpost.com/world/2019/02/22/india-orders-staggering-eviction-million-indigenous-people-some-environmentalists-are-cheering/?noredirect=on&utm_term=.ddbab29354b3.

Maurice, Thomas. *The History of Hindostan: Its Arts and Its Sciences, as Connected with the History of the Other Great Empires of Asia during the Most Ancient Periods of the World*, London: R. Faulder, 1795.

Maurice, Thomas. *Indian Antiquities: or, dissertations, relative to the ancient geographical divisions, the pure system of primeval theology, the grand code of civil laws, the original form of government, and the various and profound literature, of Hindostan. Compared, throughout, with the religion, laws, government, and literature, of Persia, Egypt, and Greece.* London: W. Richardson, 1796.

Mawani, Renisa. "Genealogies of the Land: Aboriginality, Law, and Territory in Vancouver's Stanley Park." *Social & Legal Studies* 14 (2005): 315–39.

Mayor, Adrienne. *The First Fossil Hunters: Paleontology in Greek and Roman Times.* Princeton, NJ: Princeton University Press, 2000.

McGowan-Hartmann, John. "Shadow of the Dragon: The Convergence of Myth and Science in Nineteenth Century Paleontological Imagery." *Journal of Social History* 47 (2013): 47–70.

McGrath, Ann, and Mary Anne Jebb. *Long History, Deep Time: Deepening Histories of Place.* Acton: Australian National University Press, 2015.

McLoughlin, Stephen. "*Glossopteris*—Insights into the Architecture and Relationships of an Iconic Permian Gondwanan Plant." *Journal of Botanical Society of Bengal* 65 (2011): 1–14.

McNamee, Gregory, ed. *The Girl Who Made Stars and Other Bushman Stories Collected by Wilhelm Bleek and Lucy C. Lloyd.* Einsiedeln: Daimon Verlag, 2015.

McPhee, John. *Basin and Range.* New York: Farrar, Straus and Giroux, 1980.

Meadows Taylor, Philip. "Ancient Remains at the Village of Jiwargi near Farozabad on the Bhima." *Journal of the Bombay Branch of the Royal Asiatic Society* 3, part 1 (1851): 179–93.

Meadows Taylor, Philip. *Descriptions of Cairns, Cromlechs, Kistvaens, and other Celtic, Druidical or Scythian Monuments in the Dekhan.* Dublin: M. H. Gill, 1865.

Medlicott, H. B., and W. T. Blanford. *A Manual of the Geology of India.* Calcutta: Office of the Superintendent of Government Printing, 1879.

Medlicott, J. G. "On the Geological Structure of the central Portion of the Nerbudda District." *Memoirs of the Geological Survey of India* 2 (1860): 197–223.

Mehta, B. H. *Gonds of the Central Indian Highlands: A Study of the Dynamics of Gond Society*. New Delhi: Concept, 1984.

Merivirta-Chakrabarti, K. R. "Reclaiming India's History-Myth, History and Historiography in Shashi Tharoor's the Great Indian Novel." *Ennen ja Nyt: Historian Tietosanomat* (September 9, 2007). Accessed July 23, 2019, http://www.ennenjanyt.net/2007/09/reclaiming-india%E2%80%99s-history-%E2%80%93-myth-history-and-historiography-in-shashi-tharoor%E2%80%99s-the-great-indian-novel/.

Mitra, Panchanan. *Prehistoric India: Its Place in the World's Cultures*. Calcutta: Calcutta University, 1923.

Monroe, James, and Reed Wicande. *The Changing Earth: Exploring Geology and Evolution*. Minneapolis, MN: West, 1994.

Moor, Edward. *Hindu Pantheon*. London: J. Johnson, 1810.

Moor, Edward. *Oriental Fragments*. London: Smith, Elder, 1834, 85–86.

Moorcroft, William. "A Journey to Lake Mánasaróvara in Ún Dés, a Province of Little Tibet." *Asiatick Researches* 12 (1816): 375–534.

Moore, Jason W., ed. *Anthropocene or Capitalocene? Nature, History, and the Crisis of Capitalism*. Oakland, CA: PM Press, 2016.

Moore, Jason W. *Capitalism in the Web of Life: Ecology and the Accumulation of Capital*. London: Verso, 2015.

Moore, Jason W. "The Capitalocene, Part I: On the Nature and Origins of Our Ecological Crisis." *Journal of Peasant Studies* 44 (2017): 594–630.

Mukharji, Projit Bihari. *Doctoring Traditions: Ayurveda, Small Technologies, and Braided Sciences*. Chicago: University of Chicago Press, 2016.

Müller, Friedrich Max. "On the relation of the Bengali to the Arian and aboriginal languages of India." In *Report of Seventeenth Meeting of the British Association for the Advancement of Science held at Oxford in June 1847*, 319–50. London: John Murray, 1848.

Müller, Friedrich Max. *Sacred Books of the East*. Oxford: Clarendon Press, 1891.

Murchison, Charles, ed. *Palæontological Memoirs and Notes of the Late Hugh Falconer*. 2 vols. London: Robert Hardwicke, 1868.

Murthy, S. R. N. "The Vedic River Saraswati: A Myth or Fact—a Geological Approach." *Indian Journal of History of Science* 15 (1980): 189–92.

Muthahar Saqaf, Syed. "A Rock Solid Project That Has Survived 2,000 Years." *Hindu* (March 9, 2013). Accessed July 23, 2019, http://www.thehindu.com/news/cities/Tiruchirapalli/a-rock-solid-project-that-has-survived-2000-years/article4491152.ece.

Myllyntaus, Timo, and Mikko Saikku, eds. *Encountering the Past in Nature: Essays in Environmental History*. Helsinki: Helsinki University Press, 1999.

Naim, C. M. "Syed Ahmad and His Two Books Called 'Asar-al-Sanadid.' " *Modern Asian Studies* 45 (2011): 669–708.

Nair, Savithri Preetha. " 'Eyes and No Eyes': Siwalik Fossil Collecting and the Crafting of Indian Palaeontology (1830–1847)." *Science in Context* 18 (2005): 359–92.

Nandy, Ashis. "History's Forgotten Doubles." *History and Theory* 34 (1995): 44–66.

Napier, J. R., and J. S. Weiner. "Olduvai Gorge and Human Origins." *Antiquity* 36 (1962): 41–47.

Narrative of a Journey from Caunpoor to the Boorendo Pass in the Himalaya Mountains . . . by Major Sir William Lloyd. And a Captain Alexander Gerard's Account of an Attempt to Penetrate by Bekhur to Gardoo, edited by George Lloyd. London: J. Madden, 1840.

Nasr, Syed. *Religion and the Order of Nature*. New York: Oxford University Press, 1996.

Neubauer, Franz. "Gondwana-Land Goes Europe." *Austrian Journal of Earth Sciences* 107 (2014): 147–55.

Newbold, Thomas John. "Summary of the Geology of Southern India." *Journal of the Royal Asiatic Society of Great Britain and Ireland* 8 (1846): 213–70.

Newton, C. T. "On the Study of Archaeology." *Archaeological Journal* 8 (1851): 1–26.

Nilsson, Sven. *The Primitive Inhabitants of Scandinavia*. Translated by John Lubbock. London: Longmans Green, 1868.

Nite, Dhiraj Kumar. "Worshipping the Colliery-Goddess." *Contributions to Indian Sociology* 50 (2016): 163–86.

"Notes on the lost river of the Indian Desert." *Calcutta Review* 59 (1874): 1–27.

Oakley, Kenneth. "Folklore of Fossils, Part I." *Antiquity* 39 (1965): 9–16.

Ogborn, Miles. "The Relations between Geography and History: Work in Historical Geography in 1997." *Progress in Human Geography* 23 (1999): 97–108.

Oldham, C. F. "The Saraswatī and the Lost River of the Indian Desert." *Journal of the Royal Asiatic Society of Great Britain and Ireland* (January 1893): 49–76.

Oldham, R. D. "On Probable Changes in the Geography of the Punjab and its Rivers: An Historico-Geographical Study." *Journal of the Asiatic Society of Bengal* 55 (1886): 322–43.

Oppert, Gustav. *On the Original Inhabitants of Bharatavarsa or India*. London: Archibald, Constable, 1893.

Oreskes, Naomi. *The Rejection of Continental Drift: Theory and Method in American Earth Science*. New York: Oxford University Press, 1999.

Oreskes, Naomi. "The Scientific Consensus on Climate Change: How Do We Know We're Not Wrong?" In *Climate Change: What It Means for Us, Our Children, and Our Grandchildren*, edited by Joseph F. C. Dimento and Pamela Doughman, 105–48. Cambridge, MA: MIT Press, 2007/2014.

Orlebar, Arthur Bedford. "Some Observations on the Geology of the Egyptian Desert." *Journal of the Bombay Branch of the Royal Asiatic Society* 2 (1845): 229–50.

O'Rourke, J. E. "A Comparison of James Hutton's Principles of Knowledge and Theory of the Earth." *Isis* 69 (1978): 4–20.

Owen, Richard. *Geology and Inhabitants of the Ancient World*. London: Bradbury and Evans, 1854.

Padel, Felix, and Samarendra Das. *Out of This Earth: East India Adivasis and the Aluminium Cartel*. New Delhi: Orient BlackSwan, 2010.

Panigrahy, K. K., G. K. Panigrahy, S. K. Dash, and S. N. Padhy. "Ethnobiological Analysis from Myth to Science: III. The Doctrine of Incarnation and Its Evolutionary Significance." *Journal of Human Ecology* 13 (2002): 181–90.

Parrinder, Geoffrey. *Avatar and Incarnation: The Divine in Human Form in the World's Religions*. Oxford: One world, 1997.

Patankar, Mayuri. "'Gondwana'/'Gondwanaland' as a Homeland of the Gonds: Storytelling in the Paintings of Gond Pilgrims." *Summerhill* 22 (2016): 39–48.

Patterson, Jessica. "Enlightenment, Empire and Deism: Interpretations of the 'Hindoo Religion' in the Work of East India 'Company Men,' 1760–1790." PhD diss., University of Manchester, 2017.

Pavgee, Narayan Bhavanrao. *The Vedic Fathers of Geology*. Poona: Arya-Bhushan Press, 1912.

Pels, Peter. "The Rise and Fall of the Indian Aborigines: Orientalism, Anglicism and the Emergence of an Ethnology of India." In *Colonial Subjects: Essays in the Practical History of Anthropology*, edited by Peter Pels and O. Salemink, 82–116. Ann Arbor: University of Michigan Press, 1999.

Peterson, Nicolas. "Studying Man and Man's Nature: A History of the Institutionalisation of Aboriginal Anthropology." In *The Wentworth Lectures: Honouring Fifty Years of Australian Indigenous Studies*, edited by Robert Tonkinson, 102–24. Canberra: Aboriginal Studies Press, 2015.

Petraglia, M. D., and Bridget Allchin, eds. *The Evolution and History of Human Populations in South Asia: Inter-disciplinary Studies in Archaeology, Biological Anthropology, Linguistics, and Genetics*. Vertebrate Paleobiology and Paleoanthropology Series. Dordrecht: Springer, 2007.

Piccardi, Luigi, and W. Bruce Masse, eds. *Myth and Geology*. London: Geological Society, 2007.

Picart, Bernard. *The Ceremonies and Religious Customs of the Various Nations of the Known World*. Vol. 3, *The Ceremonies of the Idolatrous nations*. London: Nicholas Prevost and Comp., 1731.

Piddington, Henry. "Memorandum on an Unknown Forest race (of Indian Veddas?) inhabiting the Jungles South of Palmow." *Journal of the Asiatic Society of Bengal* 24 (1855): 207–10.

Pilgrim, Guy E. "Correlation of Siwaliks with Mammal Horizons of Europe." *Records of the Geological Survey of India* 43 (1913): 264–326.

Pilgrim, Guy E. "New Siwalik Primates and Their Bearing on the Question of the Evolution of Man and the Anthropoidea." *Records of the Geological Survey of India* 45, no. 1 (1915): 1–3.

Pilgrim, Guy E. "Preliminary Note on a Revised Classification of the Tertiary Freshwater Deposits of India." *Records of the Geological Survey of India* 40 (1910): 185–205.

Pilgrim, Guy E. "Suggestions concerning the History of the Drainage of Northern India, Arising out of a Study of the Siwalik Boulder Conglomerate." *Journal of the Asiatic Society of Bengal*, n.s., 15 (1919): 81–99.

Pirsson, Louis Valentine. *Introductory Geology*. Part 1, *Physical Geology*, by L. V. Pirsson; Part 2, *Outlines of Historical Geology*, by Charles Schuchert. New York: J. Wiley & Sons, 1924.

Porter, Roy. "Gentlemen and Geology: The Emergence of a Scientific Career." *Historical Journal* 21(1978): 809–36.

Prasad, Akash K. "Gondwana Movement in Post-colonial India: Exploring Paradigms of Assertion, Self-Determination and Statehood." *Journal of Tribal Intellectual Collective India*, no. 4.3.1 (September 2017): 37–61.

Prasad, Archana. "Military Conflict and Forests in Central Provinces, India: Gonds and the Gondwana Region in Precolonial History." *Environment and History* 5 (1999): 361–75.

Pratt, J. H. *Figure of the Earth*. London: Macmillan, 1860.

Prestwich, Joseph. *Geology: Chemical, Physical, and Stratigraphical*. 2 vols. Oxford: Clarendon Press, 1886–88.

Prinsep, James. "Bactrian and Indo-Scythic Coins—Continued." *Journal of the Asiatic Society of Bengal* 2 (1833): 405–16.

Prinsep, James. "Note on the Coins, found by captain Cautley at Behat." *Journal of the Asiatic Society of Bengal* 3 (1834): 227–31.

Prinsep, James. "Note on the Proceedings." *Journal of the Asiatic Society of Bengal* 4 (1835): 500–506.

Prinsep, James. "Occurrence of the Bones of Man in the Fossil State." *Journal of the Asiatic Society of Bengal* 2 (1833): 632–35.

Prinsep, James. "On the Ancient Roman Coins in the Cabinet of the Asiatic Society." *Journal of the Asiatic Society of Bengal* 1 (1832): 392–408.

Prinsep, James. "On the Connection of Various Ancient Hindu coins with the Grecian or Indo-Scythic Series." *Journal of the Asiatic Society of Bengal* 4 (1835): 621–43.

Pymm, Rachel. "'Serpent Stones': Myth and Medical Application." *Geological Society, London, Special Publications* 452 (December 19, 2016): 163–80.

"Question Raised in Lok Sabha on Forest Right Act." India Environmental Portal, July 18, 2016. Accessed July 24, 2019, http://www.indiaenvironmentportal.org.in/content/432472/question-raised-in-lok-sabha-on-forest-right-act-18072016/.

Qureshi, Sadiah. *Peoples on Parade: Exhibitions, Empire, and Anthropology in Nineteenth-century Britain*. Chicago: University of Chicago Press, 2011.

Radhakrishna, Meena. "Of Apes and Ancestors: Evolutionary Science and Colonial Ethnography." *Indian Historical Review* 33 (2006): 1–23.

Raj, Kapil. *Relocating Modern Science: Circulation and the Construction of Knowledge in South Asia and Europe, 1650–1900*. Basingstoke: Palgrave Macmillan, 2007.

Ramaswamy, Sumathi. *The Lost Land of Lemuria: Fabulous Geographies, Catastrophic Histories.* Berkeley: University of California Press, 2004.

Ramaswamy, Sumathi. *Passions of the Tongue: Language Devotion in Tamil India, 1891–1970.* Berkeley: University of California Press, 1997.

Ramaswamy, Sumathi. "Remains of the Race: Archaeology, Nationalism, and the Yearning for Civilisation in the Indus Valley." *Indian Economic & Social History Review* 38 (June 2001): 105–45.

Rao, G. N. "Canal Irrigation and Agrarian Change in Colonial Andhra: A Study of Godavari District, c.1850–1890." *Indian Economic & Social History Review* 25 (1988): 25–60.

Rao, Ramachandra. *Śālagrāma-kosha.* Vol.1. Bangalore: Kalpatharu Research Academy, 1996.

Rappenglück, Michael A. "The Whole World Put between Two Shells: The Cosmic Symbolism of Tortoises and Turtles." *Mediterranean Archaeology and Archaeometry* 6 (2006): 223–30.

"Remarks on the Preceding Essay by the President." *Asiatick Researches* 3 (1792): 463–68.

"Remarks on two flints from Jubbulpore, Central India, exhibited at a meeting of the Society on 19th December 1867, and on the flint implements discovered there by the late Lieut. Downing Swiney." *Transactions of the Edinburgh Geological Society* 1 (1868): 199.

Rennell, James. *Memoir of a Map of Hindoostan: Or the Mogul Empire: With an Introduction, Illustrative of the Geography . . . and a Map of the Countries Situated between the Heads of the Indian Rivers, and the Caspian Sea.* London: W. Bulmer, 1792.

Report of the Ethnological Committee on papers laid before them and upon examination of specimens of aboriginal tribes brought to the Jubbulpore exhibition of 1866–67. Nagpore: M. Lawlor, 1868.

"Review: *On the Occurrence of Stone Implements in Lateritic Formations in various parts of Madras and North Arcot Districts.* By Foote R. Bruce, Geological Survey of India, Madras, 1865; 42." *Geological Magazine* 2 (1865): 503–4.

Ribeiro, Guilherme. "La genèse de la géohistoire chez Fernand Braudel: Un chapitre de l'histoire de la pensée géographique." *Annales de Géographie* 4 (2012): 329–46.

Rivett-Carnac, J. H. "Flint Implements from Central India" In *Archaeological notes on Ancient Sculpturings on rocks in Kumaon, India, similar to those found on monoliths and rocks in Europe: with other papers*, 16–28. Calcutta: Baptist Mission Press, 1879.

Rivett-Carnac, J. H. "On Stone Implements from the North Western Provinces of India." *Journal of the Asiatic Society of Bengal* 52 (1883): 221–30.

Rivett-Carnac, J. H. "Prehistoric Remains in Central India." *Journal of the Asiatic Society of Bengal* 48 (1879): 1–16.

Roberts, Janine. *From Massacres to Mining: The Colonization of Aboriginal Australia.* London: War on Want, 1978.

Robin, Libby. "Perceptions of Place and Deep Time in the Australian Desert." In *Thinking through the Environment: Green Approaches to Global History*, edited by Timo Myllyntaus, 81–99. Cambridge: White Horse Press, 2011.

Robinson, Thomas. *The Anatomy of the Earth.* London: J. Newton, 1694.

Rohit, Negi. " 'You Cannot Make a Camel Drink Water': Capital, Geo-history and Contestations in the Zambian Copperbelt." *Geoforum* 45 (2012): 240–47.

Rose, Deborah Bird. *Dingo Makes Us Human: Life and Land in an Aboriginal Australian Culture.* Cambridge: Cambridge University Press, 1992.

Rosenberg, Charles. "Toward an Ecology of Knowledge: On Discipline, Context, and History." In Rosenberg, *No Other Gods: On Science and American Social Thought*, 225–39. London: Johns Hopkins University Press, 1979/1997.

Rossi, Paolo. *The Dark Abyss of Time: The History of the Earth & the History of Nations from Hooke to Vico.* Chicago: University of Chicago Press, 1984.

Roy, Binodbihari. *Prithibir Puratattva* [The facts of the ancient earth]. Calcutta: India Press, 1914.

Royle, John F. *An Essay on the Antiquity of Hindoo Medicine: Including an Introductory Lecture to the Course of Materia Medica and Therapeutics, Delivered at King's College*. London: W. H. Allen, 1837.

Royle, John F. *Illustrations of the Botany and Other Branches of the Natural History of the Himalayan Mountains, and Flora of Cashmere*. 2 vols. London: Wm. H. Allen, 1839.

Rudwick, Martin J. S. *Bursting the Limits of Time: The Reconstruction of Geohistory in the Age of Revolution*. Chicago: University of Chicago Press, 2005.

Rudwick, Martin J. S. *Earth's Deep History: How It Was Discovered and Why It Matters*. Chicago: University of Chicago Press, 2014.

Rudwick, Martin J. S. *Georges Cuvier, Fossil Bones, and Geological Catastrophes: New Translations & Interpretations of the Primary Texts*. Chicago: University of Chicago Press, 1997.

Rudwick, Martin J. S. *The Great Devonian Controversy: The Shaping of Scientific Knowledge among Gentlemanly Specialists*. Chicago: University of Chicago Press, 1985.

Rudwick, Martin J. S. *Worlds before Adam: The Reconstruction of Geohistory in the Age of Reform*. Chicago: University of Chicago Press, 2008.

Rupert Jones, T. "Miocene Man in India." *Natural Science* 5 (1894): 345–49.

Rupke, Nicolaas A. "Eurocentric Ideology of Continental Drift." *History of Science* 34, no. 3 (1996): 251–72.

Rupke, Nicolaas A. *The Great Chain of History: William Buckland and the English School of Geology, 1814–1849*. Oxford: Clarendon Press, 1983.

Rupke, Nicolaas A. "Paradise and the Notion of a World Centre: From the Physico-theologians to the Humboldtians." In *Phantastische Lebensrdume, Phantome und Phantasmen*, edited by H.-K. Schmutz, 77–87. Marburg: Basilisken-Presse, 1997.

Russell, Robert V. *The Tribes and Castes of the Central Provinces of India*. London: Macmillan, 1916.

Sachau, Edward. *Alberuni's India: An Account of the Religion, Philosophy, Literature, Geography, Chronology, Astronomy, Customs, Laws and Astrology of India about A.D. 1030; An English Edition with Notes and Indices*. 2 vols. London: K. Paul, Trench, Trübner, 1910.

Sahlins, Marshall D. "The Origin of Society." *Scientific American* 203 (1960): 76–88.

Sahlins, Marshall D. "The Social Life of Monkeys, Apes and Primitive Man." *Human Biology* 31 (1959): 54–73.

Sahni, Birbal. "The Himalayan Uplift since the Advent of Man: Its Culthistorical Significance." *Current Science* (August 1936): 57–61.

Sahni, Birbal. *Revisions of Indian Fossil Plants*. Calcutta: Government of India Central Publication Branch, 1928.

Said, Edward W. "Invention, Memory, and Place." *Critical Inquiry* 26 (2000): 175–92.

Said, Edward W. *Orientalism*. New York: Pantheon Books, 1978.

Sandes, E. W. C. *The Military Engineer in India*. 2 vols. Chatham: Institution of Royal Engineers, 1933–35.

Sangwan, Satpal. "From Natural History to History of Nature, Redefining the Environmental History of India." *Nature and Environment* 3 (1995): 175–94.

Sangwan, Satpal. "Reordering the Earth: The Emergence of Geology as a Scientific Discipline in Colonial India." *Indian Economic and Social History Review* 31 (1994): 224–33.

Sangwan, Satpal. *Science, Technology and Colonisation: An Indian Experience, 1757–1857*. Delhi: Anamika Prakashan, 1991.

Sankhyan, A. R., L. N. Dewangan, Sheuli Chakraborty, Suvendu Kundu, Shashi Prabha, Rana Chakravarty, and G. L. Badam. "New Human Fossils and Associated Findings from the Central Narmada Valley, India." *Current Science* 103 (2012): 1461–69.

Saravanan, Velayutham. *Environmental History and Tribals in Modern India*. Singapore: Palgrave Macmillan, 2018.

Saravanan, Velayutham. "Political Economy of the Recognition of Forest Rights Act, 2006: Conflict between Environment and Tribal Development." *South Asia Research* 29 (2009): 199–221.

Scafi, Alessandro. *Mapping Paradise: A History of Heaven on Earth*. London: British Library, 2006.

Schaffer, Simon. "The Asiatic Enlightenments of British Astronomy," In *The Brokered World: Go-Betweens and Global Intelligence, 1770–1820*, edited by Schaffer, Lissa Roberts, Kapil Raj, and James Delbourgo, 49–104. Sagamore Beach, MA: Science History, 2009.

Schama, Simon. *Landscape and Memory*. London: Harper Collins, 1995.

Sclater, P. L. "The Mammals of Madagascar." *Quarterly Journal of Science* 1 (1864): 213–19.

Schnapp, Alain, with Lothar von Falkenhausen, Peter N. Miller, and Tim Murray, eds. *World Antiquarianism: Comparative Perspectives*. Los Angeles: Getty Research Institute, 2014.

Scott, James C. *Against the Grain: A Deep History of the Earliest States*. New Haven, CT: Yale University Press, 2017.

Scott, James C. *The Art of Not Being Governed: An Anarchist History of Upland Southeast Asia*. New Haven, CT: Yale University Press, 2009.

Secord, James A. *Controversy in Victorian Geology: The Cambrian-Silurian Dispute*. Princeton, NJ: Princeton University Press, 1986.

Secord, James A. "The Discovery of a Vocation: Darwin's Early Geology." *British Journal for the History of Science* 24 (1991): 133–57.

Secord, James A. "King of Siluria: Roderick Murchison and the Imperial Theme in Nineteenth-Century British Geology." *Victorian Studies* 25 (1982): 413–42.

Sen, D., and A. K. Ghosh, eds. *Studies in Prehistory: Robert Bruce Foote Memorial Volume*. Calcutta: Firma K. L. Mukhopadhyay, 1966.

Sengupta, Indra, and Daud Ali, eds. *Knowledge Production, Pedagogy, and Institutions in Colonial India*. Delhi: Palgrave Macmillan, 2011.

Sengupta, Nirmal. "Irrigation: Traditional vs Modern." *Economic and Political Weekly* 20 (1985): 1919–38.

Sepkoski, David. *Rereading the Fossil Record: The Growth of Paleobiology as an Evolutionary Discipline*. Chicago: University of Chicago Press, 2012.

Sera-Shriar, Efram, ed. *Historicizing Humans: Deep Time, Evolution, and Race in Nineteenth-Century British Sciences*. Pittsburgh, PA: University of Pittsburgh Press, 2018.

Sethi, Nitin. "SC Eviction Order Likely to Impact 1.89 Mn Tribal, Forest-Dwelling Families." *Business Standard*. Accessed July 23, 2019. https://www.business-standard.com/article/current-affairs/sc-eviction-order-likely-to-impact-1-89-mn-tribal-forest-dwelling-families-119022101132_1.html.

Shaffer, Elinor. *"Kubla Khan" and "The Fall of Jerusalem": The Mythological School in Biblical Criticism and Secular Literature, 1770–1880*. Cambridge: Cambridge University Press, 1975.

Sharpe, Jenny. "The Violence of Light in the Land of Desire; Or, How William Jones Discovered India." *Boundary* 2, no. 20 (1993): 26–46.

Sheth, Noel. "Hindu Avatāra and Christian Incarnation: A Comparison." *Philosophy East and West* 52 (2002): 98–125.

Shortland, Michael. "Darkness Visible: Underground Culture in the Golden Age of Geology." *History of Science*, 32 (1994): 1–61.

Showers, Kate. "A History of African Soil: Perceptions, Use and Abuse." In *Soils and Societies: Perspectives from Environmental History*, edited by John R. McNeill and Verena Winiwarter, 118–76. Isle of Harris, UK: White Horse Press, 2006.

Shryock, Andrew, and Daniel Lord Smail, eds. *Deep History: The Architecture of Past and Present*. Berkeley: University of California Press, 2011.

Siddiqui, Iqtidar Husain. "Water Works and Irrigation System in India during Pre-Mughal Times." *Journal of the Economic and Social History of the Orient* 29 (1986): 52–77.

Simpson, Thomas. "Bordering and Frontier-Making in Nineteenth-Century British India." *Historical Journal* 58 (June 2015): 513–42.
Simpson, Thomas. "'Clean Out of the Map': Knowing and Doubting Space at India's High Imperial Frontiers." *History of Science* 55 (March 2017): 3–36.
Simpson, Thomas. "Modern Mountains from the Enlightenment to the Anthropocene." *Historical Journal* 62, no.1 (2019): 553–81.
Singh, Chetan. *Natural Premises: Ecology and Peasant Life in the Western Himalaya, 1800–1950*. New Delhi: Oxford University Press, 1998.
Sivaramakrishnan, K. "Unpacking Colonial Discourse: Notes on Using the Anthropology of Tribal India for an Ethnography of the State." *Yale Graduate Journal of Anthropology* 5 (1993): 57–68.
Skaria, Ajay. *Hybrid Histories: Forests, Frontiers and Wildness in Western India*. New Delhi: Oxford University Press, 2001.
Skaria, Ajay. "Shades of Wildness: Tribe, Caste and Gender in Western India." *Journal of Asian Studies* 56 (1997): 726–54.
Skaria, Ajay. "Some Aporias of History: Time, Truth and Play in Dangs, Gujarat." *Economic and Political Weekly* 34 (1999): 897–904
Slater, Ben. "Fossil Focus: Coal Swamps." *Palaeontology Online* 1 (2011): 1–9.
Sleeman, W. H. *Rambles and Recollections of an Indian Official*. London: J. Hatchard, 1844.
Smail, Daniel. "In the Grip of Sacred History." *American Historical Review* 110 (2005): 1337–61.
Smith, E. "Notes on the Specimens of the Kankar formation, and on Fossil Bones collected on the Jamna." *Journal of the Asiatic Society of Bengal* 2 (1833): 622–31.
Smith, George. *Stephen Hislop: Pioneer Missionary and Naturalist in Central India from 1844–1863*. London: John Murray, 1888.
Smith, Grafton Elliot. *The Ancient Egyptians and the origin of Civilization*. London: Harper & Brother, 1911/1923.
Smith, William. *Strata Identified by Organised Fossils*. London: W. Arding, 1816.
Somerville, Margaret. *Water in a Dry Land: Place-Learning through Art and Story*. New York: Routledge, 2013.
Sonakia, Arun, and S. Biswas. "Antiquity of the Narmada *Homo Erectus*, the Early Man of India." *Current Science* 75 (1998): 391–93.
Spillman, Deborah Shapple. *British Colonial Realism in Africa: Inalienable Objects, Contested Domains*. Basingstoke: Palgrave Macmillan, 2012.
Spry, H. H. *Modern India: With Illustrations of the Resources and Capabilities of Hindostan*. 2 vols. London: Whittaker, 1837.
Srinivas, M. N., and M. N. Panini. "The Development of Sociology and Social Anthropology in India." *Sociological Bulletin* 22 (1973): 179–215.
Stafford, Robert A. "Geological Surveys, Mineral Discoveries, and British Expansion, 1835–71." *Journal of Imperial and Commonwealth History* 12 (1984): 5–32.
Stafford, Robert A. *Scientist of Empire, Sir Roderick Murchison, Scientific Explorations and Victorian Imperialism*. Cambridge: Cambridge University Press, 1989.
Stolte, Carolien. *Philip Angel's Deex-Autaers: Vaiṣṇava Mythology from Manuscript to Book Market in the Context of the Dutch East India Company, c. 1600–1672; Dutch Sources on South Asia c. 1600–1825*. New Delhi: Manohar, 2012.
Stone, Ian. *Canal Irrigation in British India: Perspectives on Technological Change in a Peasant Economy*. Cambridge: Cambridge University Press, 1984.
Stow, George William. *The Native Races of South Africa: A History of the Intrusion of the Hottentots and Bantu into the Hunting Grounds of the Bushmen, the Aborigines of the Country*. London: Swan Sonnenschein, 1905.

Stow, George William. "On some Points in South-African Geology." *Quarterly Journal of the Geological Society of London* 27 (1871): 497–548.
Subrahmanyam, Sanjay. "Monsieur Picart and the Gentiles of India." In *Bernard Picart and the First Global Vision of Religion*, edited by Lynn Hunt, Margaret C. Jacob, and W. W. Mijnhardt, 197–214. Los Angeles: Getty Research Institute, 2010.
Suess, Eduard. "Are Great Ocean Depths Permanent?" *Natural Science* 2 (1893): 180–87.
Suess, Eduard. *The Face of the Earth (Das Antlitz der Erde).* Translated by Hertha B. C. Sollas. 2 vols. Oxford: Clarendon Press, 1904.
Suess, Eduard. "Farewell Lecture by Professor Eduard Suess on Resigning His Professorship." *Journal of Geology* 12 (1904): 264–75.
Sundar, Nandini. *The Burning Forest: India's War against the Maoists*. New Delhi: Juggernaut, 2016.
Sundar, Nandini. *Subalterns and Sovereigns: An Anthropological History of Bastar, 1854–1996*. Delhi: Oxford University Press, 2000.
Sykes, William H. "On a Fossil Fish from the Table-Land of the Deccan, in the Peninsula of India: With a Description of the Specimens." *Quarterly Journal of the Geological Society of London* 7, nos. 1–2 (1851): 272–73.
Taneja, Anand Vivek. "Ruins and the Order of Nature" In "Nature, History, and the Sacred in the Medieval Ruins of Delhi." PhD diss., Columbia University, 2013.
Temple, Richard, ed. "Note by the Editor on the Gond Songs." In *Papers related to the Aboriginal Tribes*. Part 3, i–ii.
Temple, Richard, ed. *Papers related to the Aboriginal Tribes of Central Provinces Left in Mss by the Late Revd. Stephen Hislop, Missionary of the Free Church of Scotland at Nagpore*. 4 parts. 1866.
Thangaraj, K., Gyaneshwer Chaubey, Toomas Kivisild, Alla G. Reddy, Vijay Kumar Singh, Avinash A. Rasalkar, and Lalji Singh. "Reconstructing the Origin of Andaman Islanders." *Science* 308 (2005): 996.
Thapar, Romila. "The Theory of Aryan Race and India: History and Politics." *Social Scientist* 24 (1996): 3–29.
Thornton, Edward. *A Gazetteer of the Territories under the Government of the East-India Company*. London: W. H. Allen, 1854.
Thornton, Edward. *A Gazetteer of the Territories under the Government of the East India Company and of the Native States on the Continent of India*. London: W. H. Allen, 1858.
Tickell, S. R. "List of Birds, collected in the jungles of Borabhum and Dholbhum." *Journal of the Asiatic Society of Bengal* 4 (1833): 569–83.
Tilak, Bal Gangadhar. *The Arctic Home in the Vedas, Being also a New Key to the Interpretation of Many Vedic Texts and Legends*. Poona: Tilak Bros., 1903.
Tillotson, G. H. R. *The Artificial Empire: The Indian Landscapes of William Hodges*. Richmond: Curzon, 2000.
Tod, James. "An Account of Greek, Parthian, and Hindu Medals, Found in India." *Transactions of the Royal Asiatic Society of Great Britain and Ireland* 1, no. 2 (1826): 313–42.
Tod, James. *Annals and Antiquities of Rajasthan*. Calcutta: Society for the Resuscitation of Indian Literature, (1829–32) 1902.
Tod, James. "Indo-Graecian Antiquities." Letter to the editor, Piazza Barberini, Rome, March 2, 1835, *Asiatic Journal and Monthly Register for British and Foreign India, China and Australasia* 17 (May–August 1835): 9–13.
Tomas, David. "The Production of Ethnographic Observations on the Andaman Islands." In *Colonial Situations: Essays on the Contextualization of Ethnographic Knowledge*, edited by George W. Stocking Jr., 75–108. Madison: University of Wisconsin Press, 1991.

Traill, George William. "Statistical Report on the Bhotia Mehals of Kamaon." *Asiatick Researches* 17 (1832): 1–50.

Traill, George William. "Statistical Sketch of Kamaon." *Asiatick Researches* 16 (1828): 137–234.

Trautmann, Thomas R. *Aryans and British India*. Berkley: University of California Press. 1997.

Trautmann, Thomas R. "Dravidian Kinship and the Anthropology of Lewis Henry Morgan." *Journal of the American Oriental Society* 108 (1988): 93–97.

Trautmann, Thomas R. *India: Brief History of a Civilization*. Oxford: Oxford University Press, 2011.

Trautmann, Thomas R. "Indian Time, European Time." In *Time: Histories and Ethnologies*, edited by D. Hughes and Trautmann, 167–97. Ann Arbor: University of Michigan Press, 1995.

Turner, Frank. *Between Science and Religion: The Reaction to Scientific Naturalism in Late Victorian England*. New Haven, CT: Yale University Press, 1974.

Turner, Samuel. *An Account of an Embassy to the Court of the Teshoo Lama, in Tibet: Containing a Narrative of a Journey through Bootan, and Part of Tibet*. London: W. Bulmer, 1800.

Tylor, Edward B. *Anahuac: Or, Mexico and the Mexicans, Ancient and Modern*. London: Longman, Green, and Roberts, 1861.

Tylor, Edward B. *Primitive Culture: Researches into the Development of Mythology, Philosophy, Religion, Art, and Custom*. London: J. Murray, 1871.

Tylor, Edward B. *Researches into the Early History of Mankind and the Development of Civilization*. London: John Murray, 1865.

Vaidya, K. S. *Prehistoric River Saraswati, Western India: Geological Appraisal and Social Aspects*. Springer, 2016.

van der Geer, Alexandra, Michael Dermitzakis, and John de Vos. "Fossil Folklore from India: The Siwalik Hills and the *Mahâbhârata*." *Folklore* 119 (2008): 71–92.

Vanderpoorten, Alain, S. Robbert Gradstein, Mark A. Carine, and Nicolas Devos. "The Ghosts of Gondwana and Laurasia in Modern Liverwort Distributions." *Biological Reviews* 85 (2010): 471–87.

van Sittert, Lance. "The Supernatural State: Water Divining and the Cape Underground Water Rush, 1891–1910." *Journal of Social History* 37 (2004): 915–37.

Vanzolini, Marina, and Pedro Cesarino. "Perspectivism." In "Anthropology," *Oxford Bibliographies*. Accessed July 23, 2019. https://www.oxfordbibliographies.com/view/document/obo-9780199766567/obo-9780199766567-0083.xml.

Vidal, John. "The Tribes Paying the Brutal Price of Conservation." *Guardian*, August 28, 2016. Accessed July 23, 2019, https://www.theguardian.com/global-development/2016/aug/28/exiles-human-cost-of-conservation-indigenous-peoples-eco-tourism.

Vincent, Eve, and Timothy Neale, eds. *Unstable Relations: Indigenous People and Environmentalism in Contemporary Australia*. Crawley: University of Western Australia, 2016.

Viswanath, Rupa. *The Pariah Problem: Caste, Religion, and the Social in Modern India*. New York: Columbia University Press, 2014.

Vitaliano, Dorothy. "Geomythology: The Impact of Geological Events on History and Legend with Special Reference to Atlantis." *Journal of the Folklore Institute* 5 (1968): 5–30.

Voysey, H. W. "On Some Petrified Shells, found in the Gawilgerh Range of Hills, in April, 1823." *Asiatick Researches* 18, no. 1 (1833): 187–94.

Wadia, D. N. *Geology of India for Students*. London: MacMillan, 1919.

Wadia, D. N. "The Tertiary Geosyncline of North West Punjab and the History of Quaternary Earth-Movements and Drainage of the Gangetic Trough." *Quarterly Journal of the Geological, Mining and Meteorological Society of India* 4 (1932): 69–96.

Walker, Jeremy, and Matthew Johnson. "On Mineral Sovereignty: Towards a Political Theory of Geological Power." *Energy Research & Social Science* 45 (2018): 56–66.

Ward, William. *A View of the History, Literature and Mythology of the Hindoos: Including a Minute Description of their Manners and Customs and Translations from their Principal Works.* Serampore: Mission Press, 1815.
Warde, Paul, Libby Robin, and Sverker Sörlin. *The Environment: A History of the Idea.* Baltimore: Johns Hopkins University Press, 2018.
Webb, W. S. "Memoir relative to a Survey of Kemaon, with some accounts of the principles, upon which it has been conducted." *Asiatick Researches* 13 (1820): 293–310.
Wegener, Alfred L. *The Origin of Continents and Oceans: Translated from the Third German Edition by J. G. A. Skerl.* London: Methuen, 1924.
Wescoat, James L. "From the Gardens of the Qur'an to the 'Gardens' of Lahore." *Landscape Research* 20, no. 1 (1995): 19–29.
Wescoat, James L. "On Water, Landscape, and Architecture." *Architectural Research Quarterly* 16, no. 1 (2012): 6–8.
Wessels, Michael. "The Creation of the Eland: A Close Reading of a Drakensberg San Narrative." *Critical Arts* 28 (2014): 555–68.
Whitcombe, Elizabeth. "Irrigation." In *The Cambridge Economic History of India*. Vol. 2, *C.1751–c.1970*, edited by Dharma Kumar, 686–91. Cambridge: Cambridge University Press, 1983.
Wilford, Francis. "An Essay on the Sacred Isles in the West, with other Essays connected with that Work." *Asiatick Researches* 8 (1808): 245–367.
Wilford, Francis. "An Essay on the Sacred Isles in the West, with other Essays connected with that Work. Essay V. Origin and Decline of the Christian Religion in India." *Asiatick Researches* 10 (1811): 23–157.
Wilford, Francis. "On Egypt and other Countries adjacent to the Cálí River, or Nile of Ethiopia, from the Ancient Books of the Hindus." *Asiatick Researches* 3 (1792): 295–462.
Wilford, Francis. "On Mount Caucasus." *Asiatick Researches* 6 (1798): 455–536.
Wilford, Francis. "On the ancient Geography of India." *Asiatick Researches* 14 (1822): 373–470.
Willey, G. R., and Philip Phillips. *Method and Theory in American Archaeology.* Chicago: University of Chicago Press, 1958.
Wilson, Ara. "The Sacred Geography of Bangkok's Markets." *International Journal of Urban and Regional Research* 32 (2008): 631–42.
Wilson, Daniel. *Prehistoric Man.* Cambridge: Macmillan, 1862.
Wilson, Herbert Michael. *Irrigation in India.* Washington, DC: Government Printing Office, 1903.
Wilson, Horace Hayman. *Ariana Antiqua: A Descriptive Account of the Antiquities and Coins of Afghanistan.* London: East India, 1841.
Wilson, Horace Hayman. "Lecture on the present State of the Cultivation of Oriental Literature." *Journal of the Royal Asiatic Society of Great Britain & Ireland* 13 (1852): 191–215.
Wilson, Horace Hayman. *The Vishnu Purana: A System of Hindu Mythology and Tradition, Translated from the Original Sanscrit, and Illustrated by Notes Derived Chiefly from Other Purâńas.* London: Trübner, 1840.
Wilson, Leonard G. "Brixham Cave and Sir Charles Lyell's . . . *The Antiquity of Man*: The Roots of Hugh Falconer's Attack on Lyell." *Archives of Natural History* 23 (1996): 79–97.
Withers, Charles W. J. "Geography, Enlightenment, and the Paradise Question." In Livingstone and Withers, *Geography and Enlightenment*, 67–92, 1999.
Witzel, Michael. "Moving Targets? Texts, Language, Archaeology and History in the Late Vedic and Early Buddhist Periods." *Indo-Iranian Journal* 52 (2009): 287–310.
Witzel, Michael. *The Origins of the World's Mythologies.* New York: Oxford University Press, 2013.
Witzel, Michael. "Rama's Realm: Indocentric Rewritings of Early South Asian Archaeology and History." In *Archaeological Fantasies: How Pseudoarchaeology Misrepresents the Past and Misleads the Public*, edited by Garrett G. Fagan, 203–32. London: Routledge, 2006.

Witzel, Michael. "Vedas and Upaniṣads." In *The Blackwell Companion to Hinduism*, edited by Gavin Flood, 68–101. Oxford: Blackwell, 2003.

Woodward, A. Smith. "The Antiquity of Man." *Nature*, November 6, 1919, 212–13.

Woodward, B. B., and Mark Harrison. "Royle, John Forbes (1798–1858), Surgeon and Naturalist." *Oxford Dictionary of National Biography*. September 23, 2004. Accessed August 25, 2019. https://www.oxforddnb.com/view/10.1093/ref:odnb/9780198614128.001.0001/odnb-9780198614128-e-24239.

The Works of Sir William Jones: With the life of the author by Lord Teignmouth. In thirteen volumes. London: John Stockdale, 1807.

Worster, Donald. *Dust Bowl: The Southern Plains in the 1930s*. New York: Oxford University Press, 1980.

Worster, Donald. *Nature's Economy: A History of Ecological Ideas*. Cambridge: Cambridge University Press, 1977.

Wright, Thomas. *The Celt, the Roman, and the Saxon*. London: A. Hall, Virtue, 1852.

Wujastyk, Dominik. "'A Pious Fraud': The Indian Claims for Pre-Jennerian Smallpox Vaccination." In *Studies on Indian Medical History*, edited by G. J. Meulenbeld and Dominik Wujastyk, 121–54. Delhi: Motilal Banarsidass, 2001.

Yale North India Expedition: Palaeolithic Human Industries in the Northwest Punjab and Kashmir and Their Geological Significance. New Haven, CT: Connecticut Academy of Arts and Sciences, 1934–36.

Young, Peter. *Tortoise*, London: Reaktion Books, 2003.

Young, Robert B. *The Life and Work of George William Stow: South African Geologist and Ethnologist*. Cape Town: Darter Bros, 1908.

Yule, Henry. "A Canal Act of the Emperor Akbar, with some notes and remarks on the History of the Western Jumna Canal." *Journal of the Asiatic Society of Bengal* 15 (1846): 213–23.

Yule, Henry, and A. C. Burnell. *Hobson-Jobson: A Glossary of Colloquial Anglo-Indian Words and Phrases, and of Kindred Terms, Etymological, Historical, Geographical and Discursive*. London: Murray, 1903. Accessed July 24, 2019, https://dsalsrv04.uchicago.edu/dictionaries/hobsonjobson/.

INDEX

Page numbers in *italics* refer to figures.